A Brief History of **Earth**

ALSO BY ANDREW H. KNOLL

Life on a Young Planet

A
BRIEF
HISTORY
OF
EARTH

Four Billion Years in Eight Chapters

ANDREW H. KNOLL

CUSTOM
HOUSE

HarperCollins books may be purchased for educational, business, or sales promotional use. For information, please email the Special Markets Department at SPsales@harpercollins.com.

FIRST EDITION

Designed by Lucy Albanese
Chapter opener illustrations by Todd Marshall
Background images on pages iv, xii, 8, 36, 60, 90, 112, 138, 168, and
    194 © 1xpert/adobe.stock.com
Figure 3: Illustration © Macrovector/adobe.stock.com
Figures 7 and 44 (skulls): Illustrations by Alexis Seabrook

Library of Congress Cataloging-in-Publication Data has been applied for.

ISBN 978-0-06-285391-2

21 22 23 24 25 LSC 10 9 8 7 6 5 4 3 2 1

To Marsha.
For everything.

Contents

A Brief History of **Earth**

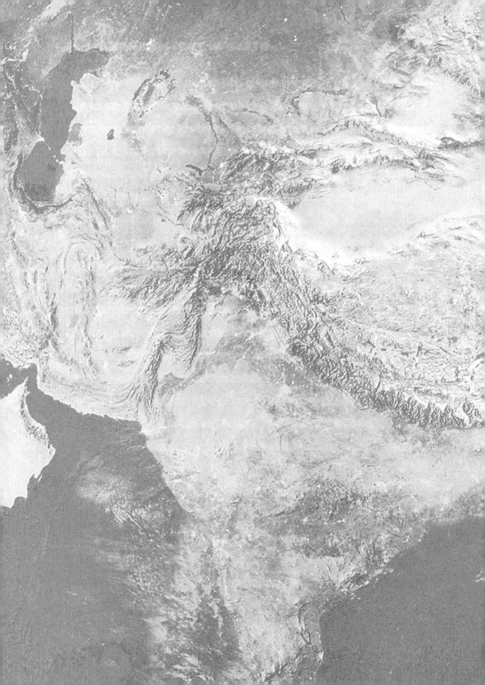

Prologue: **An Invitation**

YOU LIVE YOUR LIFE tethered by gravity to the Earth. Every step brings you in contact with rock or soil, even if hidden by a veneer of macadam or floorboards. You may think you've escaped gravity's clutches when you lift off in an airplane, but any exhilaration is fleeting; within a few hours, gravity will win, and you'll settle back onto terra firma.

Our attachment to the Earth extends well beyond gravity. The food you eat is made from carbon dioxide in the atmosphere or oceans, along with water and nutrients taken up from soil or sea. With every breath you bring oxygen-rich air into your lungs, enabling you to gain energy from your dinner. At the same time, carbon dioxide in the atmosphere keeps you from freezing. Moreover, the steel in your refrigerator door, the aluminum in your "tin" cans, the copper in your pennies, and the rare-earth metals in your smartphone all come from within the Earth. Given all this, it is remarkable how incurious most of us are about this great sphere that sustains us and occasionally, during earthquakes or hurricanes, places us in harm's way.

How can we understand Earth's place in the universe? How did the rocks, air, and water that define our existence come to be? How do we explain our continents, mountains and valleys, earthquakes and volcanoes? What controls the composition of the atmosphere or of seawater? And how did the immense diversity of life all around us come to be? Perhaps most important, how are our own actions changing both Earth and life? In part these are questions of process, but they are also historical inquires, and that's the framework of this book.

This is a story about our home, the Earth, and the organisms that spread across its surface. Everything about the Earth is dynamic, ever changing despite common but false impressions of permanence. Boston, for example, has a temperate climate, with warm summers, cold winters, and moderate precipitation distributed more or less evenly throughout the year. The seasons are predictable and if, like me, you've been around for a few decades, you may get the feeling that you've seen it all before. Meteorologists, however, will tell you that the mean annual temperature in Boston has increased by more than a degree Fahrenheit (0.6 degree Celsius) during the lifetimes of its older citizens. We also know that the amount of carbon dioxide in the atmosphere—a major regulator of surface temperature—has increased by about a third since the 1950s. Similarly, measurements tell us that global sea level is rising and the amount of oxygen dissolved in the oceans has declined by about 3 percent since the Beatles catapulted to fame.

Small changes add up through time. A plane flight from Boston to London lengthens by about one inch (2.5 centimeters) each year, as new seafloor slowly pushes North America and Europe apart. If we could run the tape backward, we'd see that 200 million years ago, New England and Old England were part of a single continent, with rift valleys like those seen today in eastern Africa just beginning to initiate an ocean basin. On the longest timescales, Earth's transformations are truly profound. For instance, free to roam on the early Earth, you would have suffocated quickly in our planet's oxygen-free air.

The story of Earth and the organisms it sustains is far grander than any Hollywood blockbuster, filled with enough plot twists to rival a bestselling thriller. More than four billion years ago, a small planet accreted out of rocky debris circling a modest young star. In its early years, Earth lived on the edge of cataclysm, bombarded by comets and meteors, while roiling magma oceans covered the surface and toxic gases choked the atmosphere. With time, however, the planet began to cool. Continents formed, only to be ripped apart and later collide, throwing up spectacular mountain ranges, most of which have been lost to time. Volcanoes a million times larger than anything ever witnessed by humans. Cycles of global glaciation. Countless lost worlds we are only beginning to piece together. Somehow on this dynamic stage, life established a foothold and eventually transformed our planet's surface, paving the way for trilobites, dinosaurs, and a species that can speak, reflect, fashion tools, and, in the end, change the world again.

Understanding Earth's history helps us appreciate how the mountains, oceans, trees, and animals around us came to be, not to mention gold, diamonds, coal, oil, and the very air we breathe. And in so doing, our planet's story provides the context needed to grasp how human activities are transforming the world in the twenty-first century. For most of its history, our home was inhospitable to humans, and indeed, among the enduring lessons of geology is a recognition of how fleeting, fragile, and precious our present moment is.

THESE DAYS, the headlines often seem to have been ripped from the book of Revelation: unprecedented wildfires in California and the Amazon aflame; record heat in Alaska and accelerating glacial melt in Greenland; giant hurricanes devastating the Caribbean and Gulf Coast, while "hundred year" floods inundate the American Midwest with increasing regularity; Chennai, India's sixth-largest city, running out of water, with Cape Town and São Paulo coming close. The news from biology is hardly better: a 30 percent decline in North American bird populations since 1970; insect populations halved; massive coral mortality along the Great Barrier Reef; rapid declines of elephants and rhinos; commercial fisheries under threat around the world. Population decline is not extinction, but it is the road down which species travel on their way to biological endgame.

Has the world run amok? In a word, yes. And we know why: the culprit is us. It is humans who pump greenhouse gases into the atmosphere, not only warming the Earth but increasing the magnitude and frequency of heat waves, drought, and storms. And it is humans who have driven species to the brink through changing land use, overexploitation, and, increasingly, climate change. With this in mind, possibly the most depressing news of all is the human response: widespread indifference, perhaps especially in my home country, the United States of America.

Why do so many people care so little in the face of planetary changes that will reshape the lives of our grandchildren? In 1968, Baba Dioum, a Senegalese forest ranger, provided a memorable answer. "In the end," he said, "we will conserve only what we love, we will love only what we understand, and we will understand only what we are taught."

This book, then, is an attempt at understanding. An invitation to appreciate the long history that has brought our planet to its present moment. An exhortation to recognize how profoundly human activities are altering a world four billion years in the making. And a challenge to do something about it.

1

Chemical **Earth**

MAKING A PLANET

IN THE BEGINNING WAS . . . well . . . a jot, a speck, a fleck at once incomprehensibly small but unimaginably dense. It wasn't a localized concentration of stuff in the vast emptiness of the universe. It *was* the universe. How it got there, no one knows.

What, if anything, came before is equally mysterious, but about 13.8 billion years ago, this primordial kernel of universe began to expand rapidly—a "Big Bang" that unleashed an immense outward tide of energy and matter. Not the rocks and minerals of our daily existence; not even the atoms from which rocks, air, and water are built. At the dawn of the universe, matter consisted of quarks, leptons, and gluons, a curious cast of subatomic particles that would eventually coalesce into atoms.

Our understanding of the universe and its history comes largely from the most ephemeral of sources: light. The luminous pinpricks that give shape to the night sky may seem unlikely history books, but two properties of light help us to understand how the universe has evolved. First, the intensity of different wavelengths in incoming radiation points to the composition of its source. Our eyes can detect only a narrow range of wavelengths, but stars and other heavenly bodies emit or absorb a broad spectrum of radiation, from radio- and micro-waves to x-rays and gamma rays, each with a story to tell. And, importantly, light obeys a strict speed limit: 299,792,458 meters per second, or 186,276 miles per second, in space. Sunlight is emitted eight minutes and twenty seconds before we see it, and for stars and other bodies farther away, the light we record emanated still earlier—much earlier for the most distant objects. That's what makes our starry sky a celestial history book.

Microwaves distributed evenly across the sky speak of the Big Bang and its immediate aftermath, and radiation from the first generation of stars, formed a few hundred thousand years after time began, is just reaching us today. How did these early stars form? It all has to do with gravity, the architect of the universe. Gravity describes the attraction between different objects, with the strength of the attraction determined by the masses of the objects and the distance between them. As atoms formed within the early, expanding universe, gravity began to pull them together. Local aggregations grew, strengthening their gravitational pull, and eventually they collapsed into hot, dense balls, so hot and so dense that hydrogen nuclei fused to form helium, releasing light and heat. When that happens, a star is born. Large, hot, and short-lived, those primordial stars set the course of all that would come later, including us.

The matter generated by the Big Bang consisted mostly of hydrogen atoms, the simplest of elements, along with some deuterium (hydrogen with an added neutron) and helium. A tiny bit of lithium formed as well, along with still smaller amounts of other light elements, but there wasn't much else. Actually, there *was* something else, but we don't quite know what it is. In the 1950s, astronomers began to use the motions of stars and galaxies (a collection of stars, gas, and dust held together by, once again, gravity) to calculate gravitational attraction in deep space, but when they summed up the mass of all known objects in the sky they found it insufficient to account for their observations. There had to be something else out there, something that interacts with normal matter through gravity but doesn't interact with light; astronomers dubbed it dark matter. Astronomers have thoughts about what dark matter might be, but no one is certain. Even more mysterious is dark energy, also deemed necessary to explain the workings of the universe. Together, dark matter and dark energy are thought to make up some 95 percent of all that exists, enigmatic constituents that we can't detect but which are thought to have played a major role in shaping the universe. We still have a lot to learn.

Let's get back to conventional matter. As the age of starlight began, the universe was a cold, diffuse cocktail of (mostly) hydrogen atoms. Early stars generated more helium, but there was nothing you could make into an Earth (see table on opposite page). Where did the iron, silicon, and oxygen needed to build our planet come from? And what about the carbon, nitrogen, phosphorus, and other elements that make up your body? These and all other elements originated in succeeding generations of stars, foundries of the atoms that would one day form our planet. At the high temperatures and pressures within large stars, light elements fused to form carbon, oxygen, silicon, and calcium; iron, gold, uranium, and other heavy elements were forged in the giant stellar explosions called supernovae. The face you see in the mirror may be decades old, but it is made of elements formed billions of years ago in ancient stars.

Through the immensity of time, stars formed and died, each cycle adding to the inventory of the elements concentrated today in Earth and life. Galaxies merged and black holes (regions so dense that no light can escape) emerged, slowly shaping the universe we observe today.

We pick up the story about 4.6 billion years ago, focusing on an unassuming cloud of hydrogen atoms, along with small amounts of gas, ice, and mineral grains within the spiraling arm of a nondescript galaxy called the Milky Way. At first, the cloud was large, diffuse, and cold (really cold, with tempera-

ELEMENTAL COMPOSITION OF THE EARTH AND LIFE
(percent, by weight)

Earth	
Iron	33
Oxygen	31
Silicon	19
Magnesium	13
Nickel	1.9
Calcium	0.9
Aluminum	0.9
Everything else	0.3
Cells in the human body:	
Oxygen	65
Carbon	18
Hydrogen	10
Nitrogen	3
Calcium	1.5
Phosphorous	1
Everything else	1.5

tures of 10–20 degrees Kelvin, or –460 to –420 degrees Fahrenheit). Probably nudged by a nearby supernova, this cloud began to collapse into a much smaller, denser, and hotter nebula. As had occurred billions of times elsewhere in the universe, gravity eventually drew most of the cloud into a hot, dense, central mass—our Sun. Most of the nebula's hydrogen went into the Sun, but ice and mineral grains were partitioned into a disk that rotated around our fledgling star, broadly reminiscent of the rings of tiny particles that encircle Saturn today (Figure 1). At first, this disk was hot enough to vaporize the minerals and ices

FIGURE 1. This remarkable image, taken by the Atacama Large Millimeter Array, shows HL Tauri, a young Sun-like star, and its protoplanetary disk. The rings and gaps evident in the image record emerging planets as they sweep their orbits clear of dust and gas. Our own solar system may have looked much like this 4.54 billion years ago. *ALMA (ESO/NAOJ/NRAO)/ NASA/ESA*

from which it formed. Over a few million years, however, it began to cool, faster in its outer reaches and slower close to the Sun's heat.

We know from our everyday experience that different substances melt or crystallize at distinct temperatures. At the Earth's surface, for example, water will turn to ice at 0 °C (32 °F), but dry ice freezes from carbon dioxide at much lower temperatures (–78.5 °C). In much the same way, the minerals found in rocks crystallize from molten precursors at temperatures that range from hundreds to more than 1,000 °C. For this reason, as the planetary disk cooled, different materials crystallized into solids at different times and distinct places, all in relation to their respective distances from the Sun's heat. Oxides of calcium, aluminum, and titanium formed first; then metallic iron, nickel and cobalt, and only later, beyond a distance from the Sun christened the frost line, ices of water, carbon dioxide, carbon monoxide, methane, and ammonia—the materials of oceans, air, and life. Bits of minerals and ice collided to form larger particles, and these coalesced into still bigger bodies. Within a few million years, only a handful of large spherical structures remained where the disk once rotated. The "third rock from the sun" was the Earth, a stony mass orbiting the Sun from a distance of about 93 million miles (150 million kilometers).

HOW, SPECIFICALLY, DID the Earth take shape, and what can we know about its infancy? If light chronicles the history of the universe, rocks tell our planet's story. When you gaze into the Grand Canyon or marvel at the peaks framing Lake Louise, you're viewing nature's library, with volumes of Earth history on display, inscribed in stone. Sediments—cobbles, sands, or muds formed by erosion of earlier rocks, or limestones precipitated from water bodies—spread across floodplains and the seafloor, recording, layer upon layer, the physical, chemical, and biological features of our planet's surface at the time and place they formed. Igneous rocks—formed from molten materials deep inside the Earth—tell us more about our planet's dynamic interior, as do metamorphic rocks forged from sedimentary or igneous precursors at elevated temperature and pressure deep within the Earth. Collectively, these rocks offer a grand narrative of Earth's development from youth to maturity, of life's evolution from bacteria to you, and—perhaps the grandest narrative of all—of the ways that the physical and biological Earth have influenced each other through time. After forty years as a geologist, I'm still amazed that cliffs along the Dorset coast of southern England allow me to conjure up a picture of the Earth as it existed 180 million years ago. Still more remarkable, as we'll see, are those rocks that tell of Earth and life billions of years ago.

If you look closely at imposing peaks in the Rocky Mountains or the Alps, another aspect of Earth history may snap into focus. Their tooth-like shapes don't reflect deposition. On the contrary, they are being sculpted by erosion, physical and chemical processes that wear away rocks, eradicating their stories. Earth writes its history with one hand and erases it with the other, and as we go further back in time, erasure gains the upper hand. Our planet coalesced some 4.54 billion years ago, but Earth's oldest known rocks date back only to about 4 billion years. Older rocks must have existed, but they've been eroded away or were buried and transformed through metamorphism into unrecognizable form. A few may still lie in some remote Canadian or Siberian hillside, waiting to be recognized, but largely, the first 600 million years of Earth history constitutes our planet's Dark Age.

How can we reconstruct Earth's infancy in the absence of historical records? It turns out we have backup copies, stored off-site, so to speak. The rocks in question are meteorites, stony survivors from the early solar system that fall to Earth from time to time. Our confidence that Earth and other planets took shape more than 4.5 billion years ago comes from geologic "clocks" trapped in the minerals that make up these special rocks. (More on dating Earth history in a bit.) Some meteorites, called chondrites, consist of rounded, millimeter-scale granules termed chondrules, thought to preserve those tiny particles that collided to form larger bodies during the earliest phases of planet formation (Figure 2). This view gains support from careful studies of chondrule composition, which includes the calcium, aluminum, and titanium minerals that were the first to

FIGURE 2. The Allende meteorite, a carbonaceous chondrite that fell to Earth in 1969. Rounded grains inside are chondrules, rocky spheroids that formed early in our solar system's history and aggregated into larger bodies, eventually to form the inner planets of our solar system, including Earth. Carbonaceous chondrites contain both water and organic molecules, furnishing materials that would eventually end up in our atmosphere, oceans, and life. The accompanying block is 1 centimeter on each side. *Matteo Chinellato (via Wiki, Creative Commons)*

condense as our solar disk began to cool, as well as rare grains ejected from a nearby supernova and later swept up as the solar system formed. Chondritic meteorites not only preserve a direct record of the early solar system, their chemical composition suggests that they are the principal materials from which Earth itself formed.

Within a few million years, most of the rock and ice around our Sun accreted into planets. In the conventional view, dust-size particles stuck together to make larger grains and these in turn aggregated into still larger bodies, eventually forming planetesimals, kilometer-scale chunks of rock like many of the asteroids found today between the orbits of Mars and Jupiter. An alternative hypothesis holds that planet-like bodies accreted directly from particles the size of pebbles. In any event, as accretion moved toward completion, only about a hundred moon- to Mars-size bodies were left. These would collide to form the planets of our solar system. One such cataclysm profoundly affected our eventual home. A few tens of millions of years after Earth had mostly accreted, a Mars-size body rammed into our infant planet, flinging rock and gas into space. Much of the ejected material eventually coalesced to form a relatively small rocky sphere, locked into permanent orbit around the Earth. The full moon may inspire poetry, but it was born in violence, its secrets unlocked through careful studies of lunar rocks.

EARTH IS A ROCKY BALL 7,920 miles (12,746 kilometers) in diameter at the equator. (Actually, our planet isn't quite spherical; because of its rotation, Earth bulges a bit at the equator and flattens toward the poles.) If you cut the Earth in half (not recommended in practice), you'll see that our planet is not homogeneous, but rather concentrically layered, like a hard-boiled egg (Figure 3). Earth's "yolk" is the core, a hot, dense inner body that accounts for about a third of our planet's mass. The core consists mostly of iron, along with some nickel and about 10 percent of lighter elements thought to include hydrogen, oxygen, sulfur, and/or nitrogen. We have to settle for "thought to include" because, with all due respect to H. G. Wells, no one has ever journeyed to the center of the Earth to retrieve a sample. Waves of energy generated by earthquakes act much like the CT scanners in hospitals, and the details of how these waves are transmitted, reflected, refracted, or absorbed within the planet reveal the core's dimensions and density. The latter requires that the core consist mostly, but not quite entirely, of iron. Laboratory experiments and calculations indicate an admixture of light elements like those noted above can account for the observed density, but the precise nature of the mixture remains unknown because no one composition provides a unique solution to the problem. The inner core—a ball 762 miles (1,226 kilometers) in diameter—is solid, while the outer core (some

1,475 miles, or 2,260 kilometers thick) remains molten and slowly moves by convection, as hotter, denser material near the base rises and cooler, less dense matter toward the top sinks. This motion of the outer core generates an electrical dynamo, resulting in Earth's magnetic field. You may not think about the magnetic field on a daily basis, but you should be grateful that it exists. The field protects our atmosphere from being stripped by solar wind (an energetic stream of charged particles emanating from the Sun) while usefully directing compasses to point (approximately) north.

Earth's mantle—the white of our planetary egg—surrounds the core. About two-thirds of our planet's mass, the mantle consists mainly of silicate minerals—minerals rich in silicon dioxide (SiO_2—quartz in its pure crystalline form) joined by magnesium and lesser amounts of iron, calcium, and aluminum. Again, much of what we know about the mantle comes from earthquake waves, illuminated by laboratory experiments. Every now and again, however, Earth obliges us by transporting bits of mantle to the surface. Diamonds are particularly notable messengers from the deep interior. Formed 100 miles (160 kilometers) or more beneath the surface, these hard lumps of pure carbon get transported to the surface by magma, the molten source of lava and other igneous rocks. Lorelei Lee insisted that diamonds are a girl's best friend, but they're the geologist's friend, as well, because diamonds commonly contain tiny inclusions of mantle material that can be studied in the lab.

The mantle is solid, but on long timescales it convects. The precise three-dimensional pattern of mantle circulation remains a topic of debate, as does the question of whether all parts of the mantle generate volcanic rocks that rise to the surface. Geologists agree, however, that the partial melting of mantle rocks has given rise to Earth's most accessible layer, the crust.

Less than 1 percent of our planet's mass, the crust—the thin shell in our egg analogy—is the only layer we can observe and sample routinely, providing a remarkable trove of knowledge. Continents are made of crust containing quartz (SiO_2) and feldspar minerals rich in sodium or potassium, typified by the granite seen in the White Mountains of New Hampshire or the Sierra Nevadas, dramatically incised in Yosemite National Park. The crust beneath the oceans is different, consisting of basaltic rocks like those that erupt from Hawaiian volcanoes; they contain calcium- or sodium-rich feldspar minerals, but no quartz. Continental crust is both thicker and less dense than the crust beneath the oceans, causing it to "float" above the ocean crust, like ice cubes in a cool drink. Indeed, it is because water at the Earth's surface accumulates in topographic lows that basaltic crust lies mainly beneath the sea.

HOW DID OUR LAYERED EARTH come to be? We might propose that Earth's concentric layers reflect the sequential accretion of distinct materials as our planet coalesced, but that idea runs afoul of many physical and chemical observations. Rather, most scientists agree that as the nascent Earth grew larger, heat from ongoing collisions and the decay of radioactive isotopes melted the planet.

ELEMENTS, ISOTOPES, AND CHEMICAL COMPOUNDS

Elements are the basic building blocks of chemical compounds, their properties determined by the number of protons and electrons they possess. Carbon, for example, bonds with other elements in distinct patterns because of its six protons and six electrons. Oxygen is distinct because it has eight protons and eight electrons. The Periodic Table of the Elements, proudly displayed in classrooms around the world, provides a systematic view of how the protons and electrons of the 118 known elements determine the chemical bonds they make, and so their distributions in nature.

All carbon atoms have six protons and six electrons, but the number of neutrons can vary. Most carbon atoms—about 99 percent—are carbon-12, which contains six neutrons and as well as six protons, for an atomic weight of 12 (where the hydrogen atom is defined as 1). About 1 percent of carbon atoms, however, have an extra neutron, for an atomic weight of 13. And a few carbon atoms in every trillion have eight neutrons, for an atomic weight of 14. Carbon-14 may be familiar because of a particular, and particularly useful, property: it is radioactive. Radioactive isotopes are unstable, breaking down through time to more stable daughter atoms. Carbon-14 spontaneously decays to nitrogen-14. In the laboratory, we can measure the rate at which this decay occurs; half of the carbon-14 present in a sample will decay to nitrogen in 5,730 years (called its half-life). This makes C-14 a valuable chronometer for archaeological research. After a few tens of thousand years, however, there simply isn't enough C-14 left in a sample to measure accurately, so we must look to other isotopes, especially those of uranium, to date Earth's deep history.

While the number of protons and electrons dictate an element's identity, and therefore the kinds of chemical reactions in which it will participate, the differing weights of isotopes influence the rates at which reactions take place, and radioactive isotopes of a number of elements provide tools for calibrating Earth history. As we'll see, these features of isotopes make them indispensable for research on the history of Earth and life.

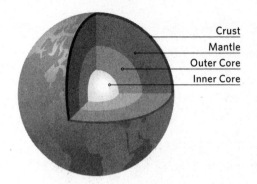

Crust
Mantle
Outer Core
Inner Core

FIGURE 3. A cross section of the Earth, showing our planet's internal zonation. The crust on which we tread is only a thin surficial veneer, and the atmosphere and oceans are even thinner.

Heavier elements, especially iron, sank to the center, while minerals of magnesium silicate and other combinations of iron, aluminum, calcium, sodium, potassium, and silica formed an outer layer. Earth's concentric structure of core and mantle emerged, with the surficial layer of crust soon to follow.

How did the crust form? To answer this question we have to return to a statement made earlier: different minerals melt or crystallize at distinct temperatures. For a few million years after the Earth formed, the hot mantle generated molten materials that rose to the surface and spread out across the planet, forming what planetary scientists call a magma ocean. If you've ever seen fresh lava streaming from Kilauea, Hawaii's most active volcano, you'll have some sense of the landscape: a rough black surface, glowing incandescent orange within cracks and fresh flows, all enveloped by a roiling layer of steam.

As heat wicked away to the atmosphere, the magma ocean soon cooled to form a primordial crust of broadly basaltic composition. And as this crust thickened and began to melt from its base, silica-rich rocks broadly similar to granite began to form—the first continental crust. A record of early crustal evolution is preserved in tiny mineral grains called zircons. The mineral zircon (zirconium silicate, $ZrSiO_4$) forms as silica-rich igneous rocks crystallize from molten magma. Zircons have a remarkable property that commends them to geologists: as they crystallize, zircons incorporate a bit of uranium into their structure. They don't take up lead because lead ions are too large to fit into the growing crystals. Why does this matter? Some uranium ions are radioactive: uranium-235 and uranium-238 decay to lead-207 and lead-206, respectively, at rates that can be measured in the laboratory. Uranium-238 has a half-life of 4.47 billion years, meaning that on this timescale half of the uranium-238 in a sample will have decayed to lead-206; similarly, uranium-235 has a half-life of 710 million years. Because no lead entered the zircons as they formed, any lead we measure in them today must have formed by the radioactive decay of uranium. So, by the careful measurement of uranium and lead in zircons, we gain a clock, Earth's best chronometer for calibrating our planet's deep history.

Okay, so zircons help us to tell geologic time, but if we have no Earth rocks older than four billion years, how can zircons shed light on our planet's earliest history? To answer that, we must go to the beach. At my family's favorite seashore, the North Shore of Massachusetts, beach sand that we shape into castles formed by the erosion of ancient highlands whose remnants can be seen in the White Mountains of New Hampshire and other ranges along the backbone of New England. The mountains expose granites that formed during a mountain-building event 400 million years ago; we know their age because the granites contain zircons that pinpoint their time of formation. Through time, some of those zircons were eroded from the mountains and swept down rivers to the coast, eventually winding up (for now) as grains in the sandy beaches of Massachusetts. The beaches, then, are modern but they are built of much older sand grains, including 400-million-year-old zircons.

This explains how zircons can illuminate Earth's Dark Age. In Western Australia, an unprepossessing outcrop of orange-weathered rocks called the Jack Hills Formation exposes sandstones and gravels deposited by rivers some three billion years ago. The age of the rock is interesting enough—we don't have many sedimentary rocks this old—but the true gift of Jack Hills becomes apparent when we look closely that the grains that were cemented into sandstone all those eons ago. Among the grains are zircons, about 5 percent of them older than four billion years. The oldest clock in at 4.38 billion years—nearly the age of the planet. Similar discoveries have recently been made in South Africa and India.

What can we learn from these ancient minerals? First, zircons don't form in all igneous rocks; most occur in silica-rich crust, with compositions along the chemical road to granites. Zircons, then, suggest that the differentiation of Earth's crust began early in our planet's history. The chemistry of oxygen in the zircons also suggests that liquid water was already present 4.38 billion years ago; Earth's hydrosphere is nearly as old as the planet. And some old zircons contain minute inclusions of other minerals that can be used to infer the properties of Earth's interior more than 4 billion years ago. Perhaps most interesting—and controversial—are tiny flecks of graphite, a mineral made of pure carbon, in a 4.1-billion-year-old zircon. Could this be a scrappy signature of life? We'll return to this question in chapter 3. For now, let's continue to examine the slowly emerging portrait of our planet in its youth.

AT THIS POINT, we've accounted for Earth's bulk composition, but what about the features most critical for life—the water in our oceans and the gas in our atmosphere? For many years, planetary scientists hypothesized that Earth's air and water came mainly from comets added to the growing Earth as a late-stage veneer. Comets, famously described as "dirty snowballs," are messengers from the outer realms of the early solar system, made mostly of ice, with only a small admixture of rocky materials. Recent progress in understanding the chemistry of comets allows us to put cometary origins hypotheses to the test; illumination comes from the isotopes of hydrogen. We have a pretty good sense of the relative amounts of hydrogen and deuterium (already introduced, an isotope of hydrogen that contains a neutron as well as a proton and electron) in water and other hydrogen-bearing substances on Earth. Because of this, plausible candidates for the source of Earth's water should have a similar ratio of hydrogen to deuterium. Comets, sadly, fail this test; their distinct hydrogen chemistry suggests that they can account for no more than about 10 percent of Earth's water.

The rest of our water, as well as the gases in our atmosphere and the carbon in our bodies, arrived in some of the meteorites that built the planet as a whole, especially certain types of chondritic meteorites thought to have arrived during late stages of Earth's growth. One group of chondrites, termed carbonaceous chondrites, deserves particular attention, as they contain 3–11 percent water by mass, mostly bound chemically into clays and other minerals, as well as about 2 percent organic matter (molecules in which carbon and hydrogen are bound together), including amino acids like those found in proteins. Chondritic meteorites, then, provide a source of water and carbon, and unlike comets, they pass the hydrogen isotope test. Thus it looks like chondritic meteorites of different kinds provided most of the rock, water, and air that we call home.

On the early Earth, heat would have driven water vapor, nitrogen gas, and carbon dioxide from the interior to form a hot, dense atmosphere, perhaps a hundred times more dense than the one we experience today. As Earth cooled, however, most of the water vapor condensed to liquid, raining out to form the oceans. At the same time, some of the atmosphere's carbon dioxide reacted with rocks and water to form limestone, returning to the solid Earth as sediments. Perhaps this Earth looked even more like Hawaii, with cloud-shrouded volcanoes emerging from the sea. It might have had an alien cast, however, as some scientists think that small organic molecules, formed by radiation-driven chemical reactions, would have formed an orangish haze in the thick early atmosphere.

Degassing was not complete—there is still more water in the mantle than there is in the oceans. Nor was the movement of water from mantle to surface a one-way street. There is reason to believe that the hot mantle of Earth's youth could hold less water than its modern counterpart, so early oceans may have been larger than today's. One thing is clear: oxygen gas was not part of this early atmosphere. As we'll see in chapter 4, the oxygen that sustains us came later, supplied by biological rather than purely physical processes.

As Earth cooled and began to differentiate, the influence of large meteors slowly waned. Meteorites do still strike the Earth. In 1992, a small meteorite crushed a car in the town of Peekskill, New York, and visitors to the magnificent Meteor Crater, near Flagstaff, Arizona, can stare into a hole nearly three-quarters of a mile (1.2 kilometers) across, excavated by an impact some 50,000 years ago. That said, the *frequency* of impacts and the maximum size of the meteorites have declined through time. On the young Earth, impacts capable of vaporizing early oceans continued for some time. Evidence for this comes not from our own planet, but from our planetary neighbor, Mars, where an ancient, cratered surface is still preserved in the southern highlands. There are some giants among these craters; a remarkable impact structure called Hellas Planitia is about 2,300 kilometers (1,400 miles) across—roughly the distance from Boston to New Orleans. The energy from such an impact would make atomic bombs seem like firecrackers.

The exact timescale of waning impacts remains a subject of vigorous debate. Since the early human exploration of the Moon, it has been popular to think in terms of a Late Heavy Bombardment, an interval circa 3.9 billion years ago characterized by particularly strong meteoritic pummeling of the inner solar system. The empirical evidence for this comes mostly from samples collected by astronauts from different parts of the lunar surface. Surprisingly, widespread samples contain evidence for shock events dated around 3.9 billion years. This was originally interpreted as a discrete spike in the rate of meteorite impact, explained by models that showed how the jostling of Saturn's and Jupiter's orbits could have expelled a lot of material from the outer solar system. Some planetary scientists, however, view the matter differently, arguing that the widespread evidence of 3.9-billion-year-old impacts on the moon actually derives from one big event, not a starfleet of separate meteors. Others argue that the apparent 3.9-billion-year-old peak is an artifact that reflects a long-term decrease in impact intensity through time. Newer models of solar system dynamics support the idea of a discrete bombardment event, but suggest that it might have happened much earlier. At the moment, many scientists believe that by 4.2–4.3 billion years ago, impacts capable of vaporizing the oceans no longer menaced the Earth.

This remarkable drama of Earth's birth—accretion from ancient star stuff, global melting and differentiation that shaped our planet's interior, the formation of oceans and atmosphere—played out on a timescale of 100 million years or less. By 4.4 billion years ago, Earth had recognizably become a rocky planet bathed by water beneath a veneer of air. Continents had started to form, but were small and may have been mostly inundated by the sea. I envision the young Earth as something like a global Indonesia, with arcs of volcanoes peaking out above the sea, but only limited continent-like landmasses. Earth was swaddled by a thick atmosphere, but it was air without oxygen; human time travelers wouldn't last long on the primitive Earth. Thus, despite some familiar features, this was not yet our Earth. The world we know of large continents, breathable air—and life—was yet to come.

2

Physical Earth

SHAPING THE PLANET

THE FLATIRONS, west of Boulder, Colorado, project upward like gigantic teeth snapping at the sky, their vertical thrust accentuated by gently undulating plains to the east. We all know the great topographic features of our planet: the Rockies, Alps, and other mountain chains contrasted with vast level stretches of prairies, steppes, and coastal plains. Continents and volcanic islands, often in chains like an incandescent necklace, emerge from a vast expanse of ocean. Earthquakes are a constant threat in some parts of the world, but barely known in others. How did the surface features of our planet come to be, and what do they tell us about Earth's inner workings?

Summarizing his own exploration of Earth's intricacies, the acclaimed author John McPhee once wrote, "If by some fiat I had to restrict all this writing to one sentence, this is the one I would choose: The summit of Mt. Everest is marine limestone." Mount Everest, with fossil shells more than five miles above the sea; the Flatirons, with originally horizontal beds now nearly vertical; Fuji-san rising dramatically above the rice fields of Honshu—these and many other features force us to view Earth's surface as dynamic, an ever-changing kaleidoscope of geography, topography, and climate. This perspective is generally appreciated today, but it was a long time coming.

For millennia, our ancestors accepted that Earth's physical features were permanent—unchanging barriers, corridors, resources, and totems that circumscribe our lives. The view of Earth as static began to crack in the seventeenth century, when Nicolas Steno, court physician to the Medici family, recognized *glossopetrae*—tongue-shaped stones that weather out of Tuscan hillsides—as the teeth of once-living sharks. Steno reasoned that as the sharks died and decayed, their teeth settled into sediments on the seafloor. If we accept this, the discovery of shark teeth in the hills above Florence meant that either the sea was once higher than it is today or the rocks that form the hills had been lifted up above the sea.

The concept of geologic impermanence gained traction more than a century later, through the writings of James Hutton, commonly considered the father of modern geology. Like other late-eighteenth-century naturalists, Hutton observed the close match between plants and their environment as he strolled the hills near his Edinburgh home. Similarly, seaweeds and sea anemones in waters of the nearby Firth of Forth seemed well suited to their own habitat. But Hutton noticed something else. Erosion was slowly but inexorably wearing down the hills. And sand and mud produced by this erosion were gradually filling in the firth.

To Hutton, this presented a conundrum. If these habitats were in a continual state of decay, how could the species they support, so manifestly designed for the environments where they thrive, persist for long intervals of time? Hutton's solution was elegant in its simplicity: through time, any given mountain will erode away, but uplift (Hutton favored heat as a mechanism) will produce a new one. Similarly, bays may fill in, but movement within the Earth will ensure that new embayments continue to form. Earth's environmental constancy, then, is maintained dynamically, through a balance between uplift and erosion.

If geologists have a mecca, it is Siccar Point, a rocky promontory along the Scottish coast east of Edinburgh. Here, flat-lying sandstones sit above an eroded surface of older rocks oriented vertically (Figure 4). The vertical rocks at the base of the exposure were deposited long ago as horizontal layers of sediment that accumulated on an ancient seafloor, one after another. Later, geologic forces thrust them upward and tilted them to their current orientation. Still later, erosion sculpted a planar surface atop the vertical beds, and that was eventually covered by new sediments deposited by rivers that crossed an ancient floodplain. And now the whole assemblage sits above

FIGURE 4. Siccar Point, in Scotland, where James Hutton grasped the dynamism of the Earth and the immensity of time. *Andrew H. Knoll*

the North Sea, slowly eroding away. Visiting by boat in 1788, Hutton recognized the dynamism he had inferred from Scottish hillsides and realized that the history apparent at Siccar Point required an immense interval of time to play out. As Hutton's companion John Playfair remembered it years later, "The mind seemed to grow giddy by looking so far into the abyss of time." Hutton had no way of knowing the ages of Siccar Point rocks, but we now understand that the vertical beds were deposited 440–430 million years ago, during the Silurian Period, while the overlying sandstones date from the Devonian Period, some sixty million years later.

As nineteenth- and early-twentieth-century geologists mapped the Earth, Hutton's repeated cycles of uplift and erosion became ever clearer. But trained eyes in places like the Alps saw that the faults and folds exposed in mountainsides required more than vertical motion. Rocks had to move laterally, as well. Our modern understanding of Earth's active surface and the features it produces began to take shape early in the twentieth century, through the writings of a German meteorologist named Alfred Wegener. Like many a youngster, captivated by a globe on rainy days, Wegener noticed that if we could close the Atlantic Ocean, the nose of Brazil would snuggle nicely into the bight of West Africa, while eastern North America would nestle cozily against the Sahara. Perhaps the continents were not fixed in place, but rather wandered across the Earth's surface, occasionally colliding and throwing up mountain ranges? Might ocean basins reflect diasporas of once-contiguous landmasses?

Wegener summarized his ideas in a 1915 book, *The Origins of Continents and Oceans*. To say the response to his hypothesis was "mixed" understates the vigor of the debate that followed. Prominent North American and European Earth scientists, retrospectively labeled "fixists," rejected Wegener's ideas because they couldn't conceive of a mechanism by which the continents could plow across ocean basins. Southern Hemisphere geologists were more enthusiastic. Not only did they recognize the geometric fit of continents discussed by Wegener, they also knew that geologic features on either side of the Atlantic Ocean suggested earlier contiguity. Fossils helped. For example, circa 290- to 252-million-year-old leaves called *Glossopteris* were known to occur in southern Africa, South America, India, and Australia—later they would also be found in Antarctica. The traditional explanation that these plants migrated from continent to continent via now-vanished land bridges struck Southern Hemisphere geologists as no less preposterous than drifting continents. Of course, the fixists held prestigious professorships in European and North American universities, trumping those poor southern souls who simply looked at the rocks.

To solve the riddle of drifting continents, scientists had to turn to the oceans. For most of human history, the deep seafloor was terra incognita. Mariners plied surface waters, but no one knew what lay below. This began to change during World War II, when sonar designed to spot enemy submarines revealed networks of mountains and trenches in the deep sea. In the 1950s, American scientists Bruce Heezen and Marie Tharp discovered the mid-Atlantic ridge, a remarkable mountain chain that bisects the Atlantic seafloor from north of Iceland (itself part of the ridge) to the tip of the Antarctic Peninsula. Similar

FIGURE 5. The revolutionary map of the Earth produced by Bruce Heezen and Marie Tharp in 1977. Long, fault-scarred mountain chains rise from the deep sea floor. *World Ocean Floor Panorama, Bruce C. Heezen and Marie Tharp, 1977. Copyright by Marie Tharp 1977/2003. Reproduced by permission of Marie Tharp Maps LLC and Lamont-Doherty Earth Observatory.*

features mark the Pacific, Indian, and Southern oceans, memorably conspicuous in Heezen and Tharp's game-changing map of the Earth, with the oceans drained (Figure 5). Emerging knowledge of the Earth beneath the sea made clear that it was time to think about our planet in new ways.

Princeton geologist Harry Hess, whose wartime observations laid the groundwork for this new understanding of ocean basins, hypothesized in 1962 that oceanic ridges play an important and specific role in the Earth system: they are the places where ocean crust is born, slowly but surely separating the continents on either side. Within a year, Hess's proposed "seafloor spreading" was confirmed by British geologists Frederick Vine and Drummond Matthews. The key was magnetism. Minerals susceptible to magnetism—like the well-named iron oxide mineral magnetite—align along Earth's magnetic field when they crystallize, recording the orientation of the field at their time and place of formation. For reasons still debated, Earth's magnetic field switches orientation by 180 degrees every few hundred thousand years. Vine and Matthews observed that the magnetic signatures of seafloor crust in the Atlantic Ocean formed a pattern of parallel stripes, delineated by magnetic field reversals over millions of years. The stripes were symmetrical about the mid-ocean ridge, and when the crustal rocks were dated using radioactive isotopes, it became clear that the youngest rocks lay closest to the ridge. Correspondingly, the seafloor grew older, stripe by stripe, as one moved away toward Europe or North America. Hess was right: new ocean crust forms at ridges, increasing the distance between Boston and my favorite London pub by about 1 inch (2.5 centimeters) each year. On a human timescale that seems negligibly slow—it certainly doesn't crimp my travel—but over the past 100 million years, the Atlantic Ocean has widened by nearly

1,600 miles (2,500 kilometers). In effect, seafloor spreading solved the problem of drifting continents, and a new paradigm, called plate tectonics, began to take shape.

Unless the Earth is getting larger (and it isn't), the formation of new crust at ocean ridges requires that older crust must be destroyed somewhere else. The graveyards for crust are subduction zones, broadly linear features where one tectonic plate sinks beneath another, returning crustal rocks to the mantle from which they originated. The Atlantic Ocean is slowly but inexorably widening, but the Pacific basin is ringed by subduction zones, marked by linear arrays of volcanoes and earthquakes, from the Aleutian Islands to Indonesia. In fact, it is the sinking of crustal slabs that pulls the ocean crust apart; new crust then forms passively at the ridges. As subducting slabs sink into the hot mantle, they begin to melt, generating volcanoes as molten material rises to the surface. Friction between slabs can lock them temporarily, but the continuing force of the sinking slab causes pressure to build, inevitably exceeding the friction. Movement resumes quickly and violently—an earthquake. Residents of Los Angeles and Tokyo take comfort in frequent small earthquakes, because these dissipate the friction between slabs. It is when things go quiet that it is time to worry.

Earth's surface, then, is an interacting mosaic of rigid plates, a "lithosphere" made up of crust and strong, solid mantle just below (Figure 6). About half the plates include continents that separate or collide as the plates on which they ride grow or subduct; the rest contain only oceanic crust. Mountain chains can form where oceanic crust sinks beneath the margin of a continent—the Andes provide an example. Or mountains can develop where two continents collide—the majestic Himalaya arose as peninsular India plowed into the underbelly of Asia. The more modestly scaled Appalachian Mountains lie far from subduction zones today, but they testify to collision between ancient continents 300 million years ago. Similarly, the swath that the Urals cut across Russia, separating Europe and Asia, reflects long-ago continental collision.

Plates can also slide past each other, neither generating new crust or subducting old; perhaps the most famous example is the San Andreas Fault, which slashes across California,

FIGURE 6. The Earth's surface consists of interlocking plates. Plates are pulled apart, and new ocean crust forms along oceanic ridge systems (shown as double lines); this causes continents to diverge from one another. Plates glide past each other along transform faults (single lines), but at convergent margins (toothed lines) they collide, with one plate subducting beneath the other. Volcanoes, earthquakes, and actively growing mountain belts are concentrated along convergent plate boundaries. *Map illustration by Nick Springer/Springer Cartographics, LLC*

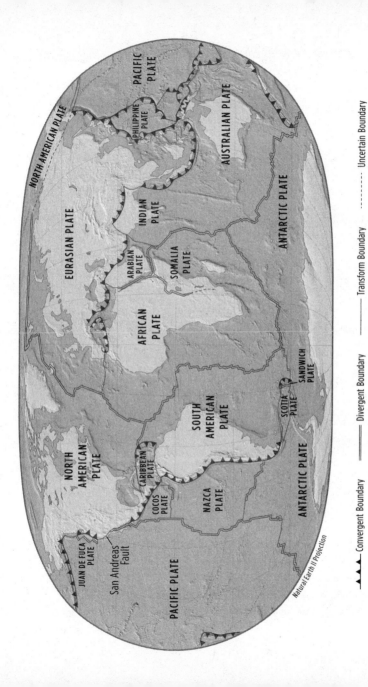

PACIFIC PLATE

JUAN DE FUCA PLATE

San Andreas Fault

NORTH AMERICAN PLATE

COCOS PLATE

CARIBBEAN PLATE

NAZCA PLATE

SOUTH AMERICAN PLATE

ANTARCTIC PLATE

SCOTIA PLATE

SANDWICH PLATE

AFRICAN PLATE

ARABIAN PLATE

SOMALIA PLATE

EURASIAN PLATE

INDIAN PLATE

ANTARCTIC PLATE

AUSTRALIAN PLATE

PHILIPPINE PLATE

PACIFIC PLATE

NORTH AMERICAN PLATE

Natural Earth II Projection

▲▲▲ Convergent Boundary   —— Divergent Boundary   —— Transform Boundary   ------- Uncertain Boundary

from north of San Francisco into Mexico. Friction between the North American plate to the east and the Pacific plate to the west generates the earthquakes that continually unsettle this region. Scientists can't stop the tremblors, but, backed by massive computing power, they are learning to predict them.

Through the work of British geophysicist Dan McKenzie and others, we now know that plate movements at the Earth's surface reflect dynamism deeper within the planet. In chapter 1, we noted that the mantle convects, with hot materials rising from its bottom and cooler ones sinking back toward the core. Ridges form where hot (and therefore, relatively buoyant) man-

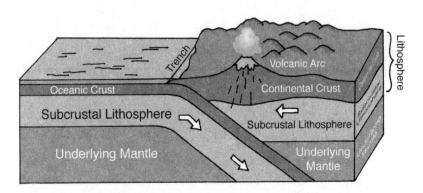

FIGURE 7. Mountains form where continents collide (e.g., the Appalachians) or oceanic crust subducts beneath a continent, as shown here (the Andes), all driven by convection in the mantle below. Trenches, linear depressions in the deep seafloor, form a surface expression of convergent plate boundaries. *Source: U.S. Geological Survey*

tle ascends toward the surface, while subduction zones coincide with descending mantle. Thus the mountains and oceans familiar to us from maps and travel reflect processes at work deep within the Earth (Figure 7).

Plate tectonics can't explain everything—it still isn't clear, for example, why one of the most powerful earthquakes ever recorded shook Missouri in 1811. That said, plate tectonics provides a compelling first order explanation for our dynamic Earth, where ocean basins form and disappear, mountains rise and eventually erode away, and earthquakes continually shatter the peace. And it has always been so—or has it?

RECONSTRUCTING EARTH'S TECTONIC HISTORY poses a geologic challenge worthy of Sherlock Holmes. We can observe and quantify spreading, subduction, and other processes in play today, but how can we know what the Earth was like ten million years ago, or two billion? For the past 180 million years or so, we have ocean crust with its magnetic stripes to guide us, enabling geologists, in essence, to play the plate tectonic tape backward. For example, if we want to know the positions of the continents ten million years ago, we can identify all oceanic crust that age or younger, (virtually) remove it, and close up the resulting gap. Seen from space, that world wasn't too different from ours, although the Atlantic Ocean was narrower and mountain chains like the Alps and Caucasus less prominent.

Fifty million years ago, the Atlantic was still smaller, and viewing that world from above, we would begin to see some unfamiliar features. Peninsular India was separated from Asia, lying to the south and surrounded by sea. Australia was just beginning to detach from Antarctica. And, with no ice sheets at the poles, a higher sea level submerged low-lying parts of Eurasia and the eastern seaboard of the United States.

One hundred million years ago, things look even more different. The Rocky Mountains are beginning to rise, but there are no Alps or Himalaya. Much of mid-continent North America and southern Eurasia is covered by shallow seas. The Atlantic Ocean is reduced to a thin band, Australia docks firmly with Antarctica, and peninsular India increasingly nestles into a geographic nook between Africa and Antarctica.

A broad pattern will have become evident: when we run the tape backward, the continents, widely dispersed today, begin to coalesce into a single great landmass. Indeed, by about 180 million years ago, we see a planet that, geographically at least, is utterly unlike our own (Figure 8). All the continents of the Southern Hemisphere unite into a single piece, called Gondwana (that's what all those *Glossopteris* leaf fossils were trying to tell us), which in turn is attached at one end to North Amer-

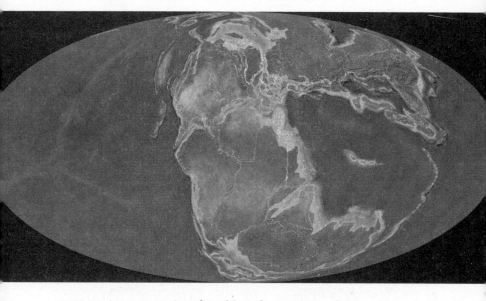

FIGURE 8. A reconstruction of Earth's surface as it existed circa 180 million years ago. The continents, which aggregated earlier, largely remain clustered. The Atlantic Ocean has just begun to open. In contrast, Tethys (the large sea south of Asia and north of the Gondwana continents) will soon close as Africa, India, and Australia separate and move northward. Eventually they will collide with Europe and Asia, producing the long mountain chain that runs from the Alps to the Himalaya, and on to New Guinea. *2016 Colorado Plateau Geosystems, Inc.*

ica and Eurasia to form a single supercontinent, Pangaea, deeply incised by a now extinct ocean called Tethys. Of course, what really happened is that about 175 million years ago, the Pangaean supercontinent began to break up, fractured by stresses from the convecting mantle below. New oceanic crust drove continental dispersal, opening new oceans, perhaps especially the Atlantic. As crust underlying the Pacific seafloor subducted beneath the westward-moving continents of North and South America, the Rockies and Andes arose. Detached segments of Gondwana moved north as the Southern Ocean opened, closing the Tethyan ocean and eventually colliding with Eurasia to form the mountainous backbone that runs from the Pyrenees to eastern Himalaya. The story continues today, as Australia's northward trek toward Asia pushes up the impressive mountains of New Guinea, their peaks reaching some 15,000 feet (4,500 meters) above sea level.

That's pretty much the tale that can be told using the seafloor record, because subduction has destroyed most oceanic crust older than 180 million years. Geology, however, indicates that plate tectonics extends much further back into Earth history. Continents resist subduction better than the seafloor, and so they preserve a far longer historical record. The dimensions and characteristics of sedimentary rock accumulations, the chemistry and distribution of granites and other igneous rocks, and the disposition of faults and folds in older mountain belts make it clear that plate tectonics has shaped our planet's surface for at least the past 2.5 billion years. Because the Earth's surface is a sphere, supercontinents that break apart will eventually reassemble, as once dispersed continents collide and reconnect. Called the Wilson Cycle, after Canadian geologist J. Tuzo Wilson, who first recognized this history, patterns of supercontinental breakup, dispersal, and reassembly have played out repeatedly through time. We have evidence that five supercontinents assembled over the last 2.5 billion years, each fated to break apart as Pangaea did. The Appalachian Mountains, the Scandinavian Caledonides, and the Urals all attest to earlier collisions between ancient continents, and Pan-African fold belts in Africa and South America record still earlier supercontinent assembly.

A PRIZED POSSESSION in my office desk is an old flip book assembled in 1979 by Chris Scotese (then a graduate student, now a world authority on Earth's ever-changing geography). Each page shows the positions of continents at a given time, and when you flip through the book rapidly, landmasses appear to move, as in an early stop-action movie. Every few seconds, words like "crash," "crunch," and "rrri-ppp" flash by, highlighting continental collisions and breakup. In 1788, James Hutton wrote that the geologic record shows "no vestige of a beginning, no prospect of an end," and that is certainly the feeling I get from Chris's flip book. But we know from chapter 1 that Earth does record vestiges of its beginning. Can we follow a trail of plate tectonic movements backward to that earliest history?

The answer is "maybe." The principal challenge of reconstructing Earth's early tectonic history is the same one encountered in chapter 1. We have few rocks older than three billion years and none to document the first 10 percent of our planet's development. The tantalizing but limited information in the chemistry and geometries of Earth's oldest preserved rocks has prompted all manner of conjecture; terms like "stagnant lid" and "sag tectonics" get bandied about, each suggesting alternatives to plate tectonics as we know it. The one point on which everyone agrees is that, early in its history, Earth's interior was hotter than it is now, and for this reason, the early lithosphere must have been thicker, but weaker than today's.

Some geologists hypothesize that as Earth's magma ocean cooled, the primordial crust cracked; magma that ascended through the fissures from the mantle below pushed the crust on either side apart, initiating the lateral movements characteristic of plate tectonics. As the crust expanded, subduction necessarily followed, and the melting of descending slabs resulted in Earth's first granite-like crust. In this view, something like plate tectonics began in Earth's infancy. In contrast, another version postulates that the first granites formed as plumes of molten magma erupted to form great piles of basalt, thickening to the point where they began to melt at their base, generating granitic rocks. Therein lies the rub. Conventionally, granite reflects the subduction and partial melting of seafloor basalts, but in this version of the story, early granites formed in the absence of plate movements. Similar debates surround other chemical details of ancient rocks, as well as the structural features of Earth's most ancient terrains. Many observations are consistent with an early onset of plate tectonics, but others underscore the uniqueness of the early Earth.

Important clues come from those ancient zircons described in chapter 1. Trace elements locked into those crystals suggest that materials transited from Earth's surface into its interior more than four billion years ago, but much more slowly than they did later on. This observation has been interpreted to indicate that on the earliest Earth, zircon-bearing magmas formed at the bottom of thick volcanic piles—a "stagnant lid" with no lateral migration or subduction. By 3.8 to 3.6 billion years ago, however, subduction had initiated, requiring some form of plate-like movements.

Another new piece of the puzzle was reported in the spring of 2020. We introduced rock magnetism earlier as key to understanding seafloor spreading and, therefore, the mechanism of plate tectonics. The logic of magnetic orientation also enables us to trace the movements of continents through geologic history. If, for example, a continent travels through time from near the equator to 30 degrees north latitude, its path can be reconstructed from the magnetic orientation of minerals in volcanic deposits erupted en route. The big question, then, is: Do the magnetic orientations of rocks formed on the early Earth document the lateral movement of landmasses? It turns out that they do. In a series of painstaking analyses, Alec Brenner, Roger Fu, and colleagues showed that more than three billion years ago an ancient terrane in what is now northwestern Australia drifted across latitudes at about the same rate that Boston is currently receding from Europe.

This strengthens the case for an early initiation of plate tectonics, although it doesn't require that the early Earth worked just like the modern. It may be that plate tectonics began regionally and coexisted with stagnant lids for some time, with episodic rather than continuous plate tectonics during Earth's infancy. In this version, convecting mantle caused incipient plates to move laterally and subduct at plate margins. Today, the pull of the subducting slab as it descends into the mantle powers plate movement, but on the early Earth, plates were so weak that they would have broken readily as subduction began, detaching the sinking slab and halting the process. Early granites could form this way, but in limited abundance. In time, as the mantle continued to cool, the lithosphere strengthened, ushering in the modern regime of plate tectonics.

EARTH'S EARLIEST TECTONIC HISTORY may remain uncertain, but many geologists argue that by about three billion years ago, plate tectonics in some reasonably modern manifestation had begun to shape our planet. The consequences were profound. Australian geologist Simon Turner and his colleagues put it neatly: "In many ways, the initiation of subduction started the clock ticking on processes that produced the Earth we know today and the environment on which we are dependent."

Plate tectonics is not an inevitable consequence of planet formation. Mars, for example, shows no evidence of ancient or modern plate movements, and neither does Venus. On Earth, however, plate tectonics was established early, setting in place the physical processes that sculpt Earth's surface and, as we'll see, maintain its environment. In consequence, Earth became more than a planet with oceans and atmosphere, mountains, and volcanoes. It became a planet capable of sustaining life.

3

Biological **Earth**

LIFE SPREADS ACROSS THE PLANET

IN EARLY 2004, the roving robot *Opportunity* touched down in Eagle crater, a small depression on the surface of Mars. I remember the night vividly, as I was watching intently from the mission's home at the Jet Propulsion Laboratory, privileged to be a member of the rover science team. Smiles, hugs, and handshakes spread rapidly when NASA announced *Opportunity*'s safe landing. Minutes later joy turned to euphoria as the rover's first pictures flickered onto our screens: *Oppy* had landed just a few meters from an outcropping of layered sedimentary rocks. Just as Earth-bound geologists had done for more than a century, we could now use the physical and chemical features of these beds to reconstruct Mars's planetary history.

Over the next few weeks, discoveries came fast and furious. The age of the rocks was, and remains, uncertain. It's not easy to construct a timeline for Martian history in the absence of well-dated volcanic rocks, but reasonable estimates suggest that the beds exposed in Eagle crater formed 3.5 to 3 billion years ago, about the age of the oldest little metamorphosed sedimentary rocks on Earth. The rocks themselves were sandstones, some exhibiting ripple marks, undulating surfaces you've probably noticed where waves wash the seashore. Ripples like those ex-posed in Eagle crater form only via transport by moving water. At the same time, chemical analysis revealed that the grains and cements making up the Eagle sandstone consist largely of salts—minerals formed, in this case, by the reaction of water with volcanic rocks. Mars, forbiddingly cold and dry today, was once relatively warm and wet.

Five weeks after the landing, NASA held a press briefing to announce this discovery. The briefing at NASA headquarters had only one rule: the scientists representing the team were to talk about water, but not life. Despite this, after an hour's de-tailed discussion of the watery fingerprints in rocks at Eagle crater, pretty much every news service on Earth rushed to post breathless pieces about life on Mars. CNN's online headline, for example, blazed: "Red Planet may have been hospitable to life." More skeptical than most was *Wired*, whose website sim-ply stated, "Mars could once support life, but did it?"

The Mars press briefing nicely illustrates what most of us—from teenagers to Nobel laureates—find most interesting about planets. It isn't the rocks; it isn't the salt, the wind, or even the water, at least not in its own right. We're transfixed by planetary exploration because planets (and, potentially, their moons) are where we may find life. Within our solar system—and at our present level of understanding, within the universe—Earth stands out as the biological planet. We don't yet know whether life ever gained a foothold elsewhere. There is at least a small chance that microbes exist today in some watery outpost of our solar system like Europa or Enceladus, icy moons of Jupiter and Saturn, respectively. Clearly however, within our planetary neighborhood, life transformed its home only on Earth. Why here? Why, to channel Humphrey Bogart, "of all the gin joints in all the towns of the world" did life arise and come to thrive on our own modest corner of the Milky Way? And how did life come to reshape the planet?

FIRST, LET'S STEP BACK and ask what it is we're trying to understand. What is life, anyway? A famous Borscht Belt joke holds that life begins when the dog dies and the kids go to college, but if we take the question seriously, what is it, really, that differentiates us—and dogs and oak trees and bacteria—from mountains and valleys, volcanoes and minerals? On the strength of our own lives, or those of our children, we might volunteer that organisms grow. True, but so do quartz crystals. But organisms not only grow, they reproduce, making more of themselves through time. Organisms harvest the energy and materials required for growth and reproduction from their environment—a set of processes that biologists call metabolism. And critically, life evolves. Once formed, a quartz crystal will not evolve into a diamond, but over billions of years, Earth's first simple organisms have given rise to a staggering diversity of species, including one with the audacity to ask how we got here.

Life, then, is characterized by growth and reproduction, metabolism, and evolution. If that reasonably circumscribes life as we know it, what might the first organisms have looked like? They certainly didn't have teeth or bones, leaves or roots. The simplest organisms alive today are bacteria and their minute cousins the archaea, tiny organisms that package everything needed for growth and reproduction, metabolism, and evolution inside a single cell. The last common ancestor of all organisms alive today must have approximated the cells of bacteria, but even the simplest bacteria are complicated molecular machines, a product of evolution, not its starting point.

For many years, the National Museum of Natural History at the Smithsonian Institution featured a gently humorous but insightful video in its early Earth gallery. The clip starred Julia Child, known to a generation of Americans as television's "French Chef." In the same delightful voice with which she guided viewers through the intricacies of *Boeuf Bourguignon,*

Julia presented a recipe for "primordial soup," the mix of simple chemicals from which life is thought to have emerged. The idea that there is a "recipe" for life is admittedly simplistic, but it gains traction when we break the complexity of organisms down into their component parts, the molecules of life.

Organisms are chemical machines that evolve through time—chemistry with a history, if you will. For this reason, laboratory exploration of life's origins focuses on how the chemical constituents of cells could have formed on a lifeless Earth. Take proteins, the structural and functional workhorses of cells. The proteins in our bodies can be large and complex, but they form by stitching together relatively simple compounds called amino acids—proteins generally contain twenty different kinds—strung together into functioning structures, much as we combine letters to make words and sentences with meaning. So, if we can synthesize amino acids, we have the building blocks of proteins. In 1953, Stanley Miller and Harold Urey demonstrated just how this have could happened on the early Earth. They filled a glass vessel with carbon dioxide (CO_2); water vapor (H_2O); methane, or natural gas (CH_4); and ammonia (NH_3), a mixture of simple molecules they thought were present in Earth's primordial atmosphere. When Miller ran a spark through the vessel to simulate the effects of lightning on

Earth, its inner wall began to turn brown. The brown staining the vessel turned out to be organic molecules—including amino acids. In a single landmark experiment, Miller and Urey showed that key building blocks of life could form by natural processes.

We can approach DNA in much the same way. DNA, the cell's instruction manual and evolutionary memory, is fiendishly complex, but it has only four distinct components, called nucleotides. The complexity—and the information—in DNA come from the linear arrangement of these nucleotides along the molecule. Like the amino acids in proteins, nucleotides in DNA form the alphabet in which DNA's information is encoded. Nucleotides, in turn, can be broken down into even simpler components: a sugar, a phosphate ion (PO_4^{3-}), and a simple organic molecule called a base. The bases can be synthesized from hydrogen cyanide (HCN) and other simple compounds likely to have been present when the Earth was young. Moreover, we've known for more than a century that sugars can be generated from simple precursors like formaldehyde (CH_2O), also thought to have been present on the ancient Earth. And phosphate ions would have been supplied by the chemical weathering of volcanic rocks. Combining these components to form nucleotides challenged scientists for decades, but in 2009,

British chemist John Sutherland and his colleagues generated two types of nucleotides under plausible early Earth conditions.

Finally, there are lipids, molecular constituents of the membranes that bound all cells. Like proteins and DNA, lipids are made of simpler units, in this case long, chain-like molecules called fatty acids, again likely generated chemically on the early Earth. Remarkably, if you splash or evaporate water that contains dispersed fatty acids, they spontaneously come together to form spheroidal microstructures that have a lot in common with the membranes that bound bacteria.

So, the principal building blocks of life, the molecules from which our cells are built, can form from natural processes under conditions plausibly found locally if not ubiquitously in our planet's infancy. It is important to underscore that this conclusion is not merely theoretical, or, for that matter, experimental. We know that reactions of the type just outlined did occur billions of years ago; their record is preserved in meteorites, those remarkable relics of our solar system *in statu nascendi*. Carbonaceous chondrites, already introduced as a source of carbon and water for the accreting Earth, contain an impressive diversity of organic molecules, including amino acids (seventy different types!), sugars, fatty acids, and more. The chemistry from which life emerged may be widespread in the universe.

So far, so good, but from here things get trickier. We know that amino acids can combine to form short linear molecules called peptides, pidgin to proteins' Shakespeare. And nucleotides can do much the same. Function and memory seem to be imminent in such molecules, but in living organisms, DNA provides the molecular instructions for protein synthesis, and proteins are needed to replicate DNA. How do we escape the chicken-and-egg dilemma of which came first?

The answer may be that neither DNA nor proteins were present in the first evolving protoorganisms. When I first studied biology in the 1970s, RNA, also built from nucleotides, was generally discussed as the cell's midwife, a series of molecules that guide the transcription of DNA into proteins, carried out within a tiny intracellular structure called a ribosome. Since then, however, the known diversity of RNA molecules has expanded amazingly, as has their documented range of function. RNA stores information, like its cousin DNA, but some RNAs act like enzymes, doing the cell's molecular work in a way once thought to be the exclusive domain of proteins. Also, small RNA molecules are now known to play a role in regulating gene expression within cells. Moreover, when biologists probed the molecular depths of the ribosome, they found that RNA lies at the functional heart of the structure. Finally, recent experiments have shown that RNA molecules synthesized in the laboratory can evolve, shaped by selection to perform specific tasks. The discovery that RNA molecules can store information, function as enzymes, and evolve leads to a big thought: perhaps the earliest entities that reproduced and evolved were made of RNA, not DNA and proteins.

The RNA World hypothesis fascinates many who study life's origins. An early RNA (or RNA-like) molecule tucked inside a spontaneously formed lipid sphere could grow, reproduce, and gradually evolve greater molecular complexity and specificity. In time, DNA would evolve from RNA precursors, providing a far more stable storehouse for the cell's information, but relinquishing other functional roles. And as amino acids interacted with RNA and DNA, proteins, which generally act much faster than RNA enzymes, would evolve to take on most of the cell's structural and functional requirements. Interestingly, recent research shows that the building blocks of both DNA and RNA could have formed under plausible prebiotic conditions, raising the possibility that the dance between DNA and RNA found in every living cell was there from life's infancy.

The challenge for RNA World hypotheses and their variants is to incorporate metabolism into the mix. Maybe the first living things were simply lipid-encapsulated RNA molecules that grew, reproduced, and evolved without any special machinery to interact with their environment. This is certainly possible, and quite a few scientists favor this idea. But if metabolism isn't necessary to generate the first life, it is, in many ways, what makes life interesting, enabling organisms to interact with the oceans and atmosphere and, eventually, transform the compositions of both. With this in mind, some scientists choose to enter the maze of life's origin through a different gate, one that emphasizes metabolism over information. In this view, the rudiments of metabolism began around energy-rich hot springs deep within ocean ridges.

Metabolism-first hypotheses have the opposite problem to that of RNA World. They provide fascinating clues to how emerging life came to interact with its environment, but attempts to evolve the information of DNA, RNA, and proteins from this base have something of an "On the sixth day . . ." feel to them. The problem of life's origins, then, remains a work in progress. What we do know is that somehow, on the primitive Earth, self-replicating, metabolizing, evolvable cells originated, setting the stage for planetary transformation. (I understand that some scientists enthuse about panspermia, the hypothesis that life on the early Earth was seeded, either physically or by aliens, from elsewhere. Perhaps microbes were blasted into space by meteors impacting Mars or some other planet, eventually coming to rest on the fertile Earth. It isn't clear that such an incubator existed in the early solar system, and seeding from an extrasolar planet, by natural means or not, would occur only at exceedingly low probability, due to both long travel times and the unlikely prospect of landing in an environment where microbial immigrants could thrive. Of course, even if we choose to entertain such ideas, they don't solve the origins problem. They simply relocate it in space and time.)

IF WE CANNOT YET fully understand how life originated, perhaps we can estimate *when* it took root on the Earth, allowing us to constrain the nature of Earth's surface when life began. Now the problem becomes geological, resting on the premise that microbial life, far older than plants and animals, can leave a tractable signature in rocks. Can organisms as tiny and seemingly fragile as bacteria leave traces that document life on the early Earth, much as dinosaur bones and petrified wood do for younger epochs?

Years ago, as a young paleontologist, I travelled to the arctic island of Spitsbergen to seek evidence of ancient microbial life. Cliffs sculpted by glaciers expose several thousand meters of sedimentary rocks deposited 850–720 million years ago (Figure 9). There are no bones or shells in these rocks, and no tracks or trails on bedding surfaces. Indeed, when these rocks formed, animals that might form fossils lay millions of years into the future. If you know how to look, however, the signature of life is written clearly in the Spitsbergen rocks.

We begin with chert, sometimes called flint—exceptionally hard rocks made of fine-grained quartz. Some readers will know the flint churches of southeastern England, faced with shiny black cobbles, the hardest rocks available to their medieval builders. To appreciate the origin of those distinctive rocks, go to the White Cliffs of Dover. These magnificent chalk bluffs contain abundant nodules of black chert, precipitated within the limey sediments as they accumulated on the seafloor some seventy million years ago. The nodules are black because they contain organic matter, trapped as the nodules grew. That's the paleontological beauty of chert; it can preserve ancient biological materials for all time, including the fossils of tiny organisms buried as the beds accumulated.

In Spitsbergen, valleys incised by glacial ice expose thick beds of limestone, some of which contain black chert nodules like those in the White Cliffs (Figure 10). When examined under the microscope, paper-thin slices of these rocks reveal a petrified microbial world, rich in beautiful, if tiny, fossils (Figures 11 and 12). Many can be identified as cyanobacteria, photosynthetic bacteria that—as we shall see—play an outsized role in Earth history. Other fossils in the chert include minute algae and protozoans, and muds deposited in beneath the shal-

FIGURE 9. A cliff made of 800- to 750-million-year-old sedimentary rocks, exposed in the glaciated highlands of Spitsbergen. These rocks, and rocks like them found globally, preserve evidence of a rich microbial biota that existed long before the evolution of plants and animals. *Andrew H. Knoll*

low sea preserve more microfossils, compressed like old Valentine's Day bouquets between layers of rock (Figure 13). The fossils in these and similarly old rocks found around the world demonstrate that animals evolved into a world already full of life—mostly microorganisms.

10.

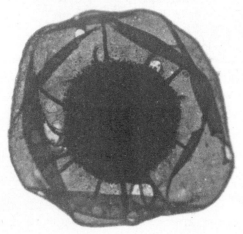

13.

FIGURES 10-13. Nodules of black chert occur within limestones in the Spitsbergen succession (Figure 10). These contain abundant and diverse microfossils of cyanobacteria (Figures 11 and 12) and other microorganisms. Mudstones in the same succession preserve beautiful fossils of single-celled eukaryotic microorganisms (Figure 13). *Andrew H. Knoll*

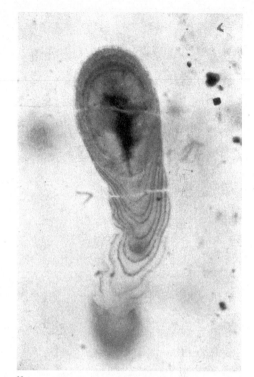

11.

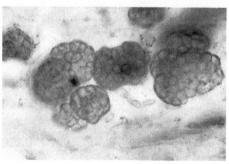

12.

Had you walked along the seashore when the Spitsbergen limestones formed, you'd have seen a seemingly unbroken coastline tinted blue green by dense mats of cyanobacteria and other microorganisms that covered the tidal zone, right to its top. Venturing offshore, you'd see more blue-green surfaces, but now projecting upward from the seafloor. These are stromatolites, fossil reefs built upward from the ancient seafloor by communities of microbes. Today, reefs are built mostly by animals, with help from skeleton-forming algae, but reefs accreted long before animals graced the Earth, telling of microbial architects. Finely layered domes, columns, and cones up to several meters thick are conspicuous in the walls of Spitsbergen cliffs (Figure 14). We can interpret them with confidence because in several corners of the modern world, where seafloor microbes are shielded from animals and seaweeds, stromatolites still form today. In these environments—as on the ancient Earth—mat-like microbial communities trap, bind, and cement sediments in place, building a rocky edifice layer by layer through time.

Chemistry reveals still more microbial signatures. Isotopes, introduced in chapter 1, are key. As we introduced earlier, carbon, the principal element of life, has two stable isotopes: Carbon-12 and Carbon-13. Carbon isotopes tell tales of ancient biology because when photosynthetic organisms fix carbon dioxide into organic molecules, they preferentially incorpo-

FIGURE 14. Stromatolites, laminated structures that formed when microbial communities trapped fine-grained sediments and bound them in place. Microbial communities colonized the firm surfaces of cobbles and then accreted upward as sediments accumulated, their growth recorded by the fine layering seen in the picture. The columns on the right are about 5 centimeters (2 inches) across. *Andrew H. Knoll*

rate CO_2 containing the lighter isotope, carbon-12, over its heavier counterpart, carbon-13. The organisms are not choosing carbon-12 with intent; it is simply that the lighter CO_2 reacts more readily with enzymes in the cell. Thus, when CO_2 is abundant, photosynthetic organisms produce organic matter that is slightly enriched in carbon-12 relative to inorganic carbon in the environment. The difference is only a few parts per thousand, but it can be measured accurately using instruments called mass spectrometers. If we go to, say, the Bahamas today and measure the carbon isotopic compositions of lime sediments and the organic matter within them, we find that the limestones and organics differ by about 25 parts per thousand. The same exercise in Spitsbergen yields similar results, indicating that a biological carbon cycle existed 850–720 million years ago. Isotopes of sulfur preserved in pyrite and gypsum similarly document an ancient sulfur cycle populated by bacteria.

Finally, ancient rocks sometimes contain actual biomolecules, produced by organisms and preserved in rocks long after their makers' deaths. Ideally, we would love to find DNA and proteins, but in really old rocks such wishes are seldom granted. The past decade has witnessed a remarkable revolution in the study of ancient DNA, but to date, the oldest DNA reliably extracted from bones or shells is less than two million years old. Similarly, proteins—such good food for bacteria and fungi—are rarely preserved in any but the youngest rocks. What do preserve are lipids, those hardy constituents of membranes. I'm fond of telling students that when they die, the last bits of them that will remain for future generations to ponder will be their cholesterol! To date, Spitsbergen rocks haven't yielded much in the way of preserved biomolecules, but other rocks of similar age preserve molecular records of diverse microorganisms. In total, then, microbial life can impart a variety of signatures to sedimentary rocks, and 850- to 720-million-year-old rocks in Spitsbergen and elsewhere preserve them in abundance.

How far back can we trace the record of life? I've worked on 1,600- to 1,500-million-year-old rocks in Australia and Siberia, and like the Spitsbergen deposits half their age, these contain microfossils, stromatolites, biomarker molecules, and isotopic evidence of microbial carbon and sulfur cycling. Doubling age yet again, we arrive at the oldest sedimentary rocks that are sufficiently well preserved to search for biological signatures: 3.5- to 3.3-billion-year-old successions preserved in remote corners of South Africa and Western Australia. These rare survivors from Earth's youth consist mostly of volcanic flows and ash, but thin intercalations of sediments enable us to ask about life's antiquity. Reports of microfossils from chert-rich rocks have proven contentious, as the simple organic microstructures in these beds may have formed by hydrothermal fluids that per-

FIGURE 15. Stromatolites in 3.45-billion-year-old sedimentary rocks from Western Australia. Along with evidence from carbon and sulfur isotopes, these structures document the presence of microbial life early in Earth's history. Scale is 15 centimeters (6 inches) long. *Andrew H. Knoll*

colated through the sediments long after deposition. Similarly, the rocks were heated during burial and tectonic deformation, destroying any biomarker molecules that might once have been present. Isotopes, however, point toward an early Earth already populated by microbes that cycled carbon and sulfur through the nascent biosphere. And stromatolites record microbial communities on the shallow seafloor (Figure 15).

Three and a half billion years ago, then, Earth was already a biological planet. And a few observations hint at still earlier life. Among the fjords of southwestern Greenland, coastal rocks include the rarest of the rare: igneous and sedimentary rocks some 3.8 billion years old. The rocks have been battered by metamorphism, and organic matter originally preserved in the sediments has been altered by heat and pressure to form graphite. The carbon isotopic composition of this material, however, looks much like that of organic matter preserved in younger rocks, suggesting a biological carbon cycle. And as mentioned in chapter 2, a tiny speck of graphite within a 4.1-billion-year-old zircon crystal from the Jack Hills in Australia is also depleted in carbon-13. We can't be sure that this most ancient carbon didn't form deep within the Earth, where it was incorporated into growing zircon crystals, but the overall message is clear. As we move backward in time, we run out of rocks to examine before we run out of evidence for life. Earth has been a biological planet for most of its long history.

WHAT DOES GEOLOGY TELL US about our planet when life began, perhaps 4 billion or more years ago? As we've already established, the youthful Earth was a watery planet, with volcanos and small continent-like masses sticking out above the waves.

Energy for prebiotic chemical reactions was widespread: Earth's surface was pummeled by ultraviolet radiation, and the decay of radioactive isotopes provided additional energetic radiation; heat from volcanoes and hydrothermal systems was ubiquitous; and lightning sliced through the early atmosphere. Hot environments existed locally, as they do today, in hydrothermal springs (think Old Faithful) and ocean ridges, but most recent data suggest an ocean and atmosphere with temperatures not too different from those we experience today.

That, in and of itself, presents a puzzle, as models for stellar evolution suggest that four billion years ago, the Sun's luminosity was only about 70 percent of its modern value. If the Sun was dim, why wasn't the early Earth an ice ball? The answer is "greenhouse gases," the bane of twenty-first-century global warming but the long-term guarantor of habitable climate. In the atmosphere, carbon dioxide, in particular, must have been present at more than one hundred times its current concentration, keeping the young Earth warm enough to maintain liquid water at its surface. The early atmosphere appears to have consisted mostly of nitrogen gas and carbon dioxide, with water vapor and variable admixtures of hydrogen gas. As noted in chapter 1, chemical observations of ancient sedimentary rocks show that oxygen gas was conspicuously absent. That's good news for the origin of life, as one thing we've learned from thousands of experiments is that when O_2 is present, prebiotic chemical reactions don't work.

So, life emerged on an Earth barely recognizable to the modern eye—lots of water and not much land, lots of carbon dioxide but little or no oxygen, hydrogen and other gases bubbling up regionally, widespread hot springs—a global Iceland, as it were. This was the anvil on which life was forged, and had you been there (don't forget to bring your own supply of oxygen) you might not have noticed the changes underfoot. But from its humble beginnings, life would expand and diversify, populating the Earth with bacteria, diatoms, sequoias, and us, shaping and reshaping our planet's surface up until the present day.

THE GEOLOGIC TIMESCALE

"Like every other branch of natural science founded on observation, we observe that the great mixed masses of the earth's crust are arranged in natural groups, and that the groups succeed in regular order." With these words, English geologist Adam Sedgwick encapsulated Earth science's great nineteenth-century revolution: the recognition of our planet's immense age and its codification in the geologic timescale.

In 1835, Sedgwick defined the Cambrian System, a succession of sedimentary rocks in Wales, distinct from other rock packages by its geometry and spatial position beneath paleontologically distinct beds of the Silurian System, proposed at about the same time by another British geologist, Sir Roderick Impey Murchison. Within a few decades, a number of systems were described and placed into a relative timescale on the basis of their stratigraphic relationships to one another. Silurian rocks were younger than Cambrian because they always lay above rocks of the Cambrian System; Devonian rocks were younger still. The time intervals during which each system was deposited became known as periods, and fossils were recognized as Earth's timekeepers. The result was the geologic timescale, or at least the part of it now known as the Phanerozoic Eon (the age of visible animal fossils).

By the time the twentieth century dawned, the relative timing of events recorded in younger parts of the rock record had become well established. But while geologists were confident that Cenozoic mammals were younger than Mesozoic dinosaurs, they didn't know the actual ages of these time intervals or their characteristic fossils. The discovery of radioactivity changed this forever. Earlier we introduced isotopes, different versions of elements distinguished by the number of neutrons they contain. Carbon has two stable isotopes, carbon-12 and carbon-13, but it also has a third isotope, carbon-14, which is radioactive. Its nucleus is unstable and breaks down through time to form nitrogen by emitting an electron (and, for those interested in the details, an electron antineutrino). We can measure the rate at which this decay occurs: the half-life of carbon-14—that is, the time it takes for half of the carbon-14 in, say, a piece of wood to decay to nitrogen—is 5,730 ± 40 years. In this way, carbon-14 provides a basis for calibrating time.

Because its half-time is relatively short, carbon-14 is useful for dating archeological materials but not the immensity of Earth history. That's a job for other radioactive isotopes, especially those of uranium. As discussed in chapter 1, zircons, formed widely in granites and related igneous rocks, are especially good for dating, enabling geologists to calibrate Earth's long history. Through a great deal of painstaking research in the field and in the laboratory, geologists have quantified geologic history. We now know not only that *Ty-*

rannosaurus rex lived during the late Cretaceous Period but that it stomped through ancient forests 68 to 66 million years ago. Radiometric dating has also been key in establishing the timetable of Earth's pre-Phanerozoic history. Figure 16 depicts the geologic timescale as we understand it in 2020; calibration of geologic time is an ongoing process, with many details still to be worked out. The figure shows not only that the fossil-rich Phanerozoic Eon has been dated with admirable resolution but also that the eon as a whole comprises only the most recent 13 percent of our planet's history. The elusive Hadean (4,540–4,000 million years), antique Archean (4,000–2,500 million), and long Proterozoic (2,500–541 million years) eons account for most of geologic time. Take a moment to appreciate the timescale; in subsequent chapters, we'll commonly use the names of eons, eras, and periods as shorthand for geologic time, just the way historians speak of the Iron Age, the Middle Ages, or the Renaissance.

FIGURE 16. The geologic timescale. Time intervals based on the International Chronostratigraphic Chart, version 2020, produced by the International Commission on Stratigraphy.

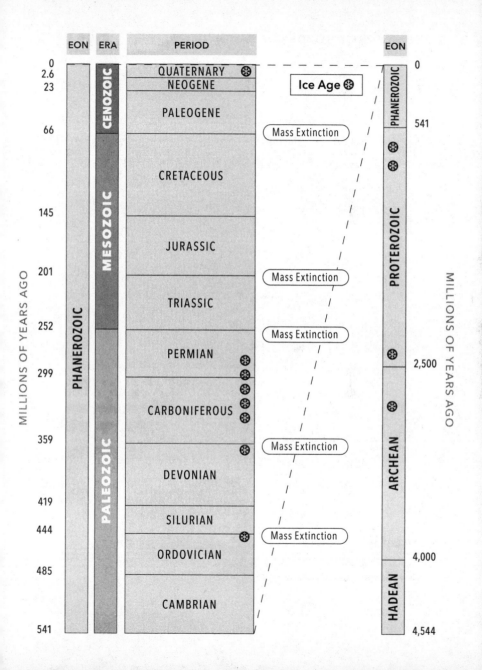

Oxygen **Earth**

THE ORIGIN OF BREATHABLE AIR

AN ATMOSPHERE WITHOUT OXYGEN? This fundamentally separates our world from that of the young Earth, but how do we know this claim is true? How can we be sure that the early Earth was so different from our own, and how can we account for the transition to a planet that can support anteaters and elephants, not to mention us? The oldest known samples of ancient atmosphere are air bubbles trapped in Antarctic ice about two million years ago, so inferences about older air and oceans must stem from chemical signatures in the rock record. Just as we learn something about Neanderthal culture from artifacts they left behind, we piece to together a picture of Earth's early atmosphere from rocks and minerals whose compositions reflect contact with ancient air and water at the time they formed.

Dales Gorge is a good place to start. A narrow canyon incised into the arid plains of northwestern Australia, the gorge exposes a thick succession of sedimentary rocks deposited nearly two and a half billion years ago (Figure 17). The rocks themselves are unusual, comprising an evenly laminated mixture of chert and iron minerals, stained russet by weathering of the iron and the red dust that permeates the Outback.

FIGURE 17. 2.5-billion-year-old iron formation exposed in Dales Gorge, Western Australia. *Andrew H. Knoll*

Appropriately, the rocks are called iron formation, and if you use a cast iron skillet in the kitchen, there's a fair chance that the pan's metal came from this type of rock.

Tellingly, iron formations do not occur on the modern sea-floor. To generate these deposits, iron must be transported through the ocean in solution, and this is only possible when O_2 is absent. Even a small amount of oxygen will react with dissolved iron, causing it to form iron oxide minerals; present-day oceans have very low iron concentrations. Iron formation, then, is a signature of oceans that were largely oxygen-free. And since surface seawater readily exchanges gases with the atmosphere, oxygen-free oceans probably sat beneath oxygen-poor air.

Iron formations are distributed widely in sedimentary basins older than about 2.4 billion years, but fall off markedly after that time, suggesting that this was when O_2 began to permeate the atmosphere and surface ocean. Other geological proxies corroborate this conclusion. For example, pyrite, or fool's gold, is probably best known to most of us as the striking golden cubes seen in museums and rock shops. Pyrite, however, helps to tell the story of oxygen. Found in ancient mudstones and some igneous rocks, fool's gold is extremely sensitive to O_2. Left to sit in a wet oxygen-rich environment, it will oxidize to sulfate, the form of sulfur found in gypsum. The timescale of this oxidation is years to decades, so although pyrite is continually eroded from rocks exposed on the continents, we essentially never see this mineral among coastal sand grains; eroded from older rocks, the pyrite reacts with oxygen and disappears.

This might seem like geological trivia, but when we examine sandstones deposited along coastlines before 2.4 billion years ago, we find grains of pyrite that were eroded from a source on land, carried downstream by rivers, and, finally, deposited along the edge of the sea—all without coming into prolonged contact with even small amounts of oxygen. In sedimentary successions younger than 2.4 billion years, we rarely see such grains. Other oxygen-sensitive minerals tell the same story.

Ancient weathering horizons strengthen the case for planetary change 2.4 billion years ago. Rocks exposed to the elements undergo chemical weathering, generating crusts of altered minerals on rock surfaces and contributing to soil. Iron comes into play once again, and for the same reasons noted earlier. When iron-bearing minerals weather beneath oxygen-free air and water, the iron they contain goes into solution and is carried away by rain and rivers. Under these conditions, when we compare the iron content of a parent rock with its weathered surface, the weathering horizon is depleted in iron. On the other hand, when oxygen is present, iron freed by weathering quickly forms iron oxide minerals, keeping it in place. Care to guess when ancient weathering horizons first show evidence of contact with O_2? The smart money is on 2.4 billion years ago, and that's the right answer.

Finally, details of sulfur isotopes in ancient pyrite and gypsum tell us that prior to 2.4 billion years ago, chemical processes in the atmosphere played a major role in Earth's sulfur cycle in a way that ceased after that time. Chemical models suggest that this telltale isotopic signature can only be imparted when oxygen levels in the atmosphere are extremely low—less than 1/100,000th of today's abundance.

FOR MORE THAN TWO BILLION YEARS, THEN—nearly the first half of our planet's history—Earth's atmosphere and oceans were essentially devoid of oxygen gas, making organisms like you and me impossible. This raises two important questions. We've already argued that Earth was a biological planet 3.5 billion years ago and perhaps much earlier. What kind of life could have prospered on this early, anoxic Earth? And, just as critical, why did this long-lasting state of the Earth's surface change 2.4 billion years ago?

The question of life without oxygen is relatively easy to address because oxygen-free environments exist today, and they teem with life. How does life persist in these forbidding (to us) habitats? In our familiar macroscopic world, plants gain energy and carbon via photosynthesis, harnessing light energy to form sugar from carbon dioxide and releasing oxygen gas as a by-product. In simplified form, the photosynthetic equation looks like this:

$$CO_2 + H_2O \rightarrow CH_2O + O_2$$

Animals do the reverse, ingesting organic molecules as food and reacting some of it with oxygen to gain energy—what we call respiration (plants also respire):

$$CH_2O + O_2 \rightarrow CO_2 + H_2O$$

The two reactions are complementary, each the reverse of the other. As a consequence, carbon and oxygen cycle back and forth between organisms and the environment, sustaining life through time.

Polish up your microscope, and you'll see that many microorganisms do the same thing—algae photosynthesize, generating organic carbon and oxygen; fungi, protozoans, and algae all respire, using up oxygen and returning carbon to the environment as CO_2. And, yes, some bacteria also cycle carbon using these pathways.

Turning carbon dioxide into sugar requires electrons, which plants and algae extract from water, generating O_2 in the process. This carries a high energy cost, but when the environment is oxygen-rich there aren't any alternatives. Where light is present but O_2 absent, however, other sources of electrons become available: hydrogen gas, hydrogen sulfide with its rotten egg smell, and iron ions in solution, among others. Under these conditions, different photosynthetic organisms take over, all of them bacterial. Not only do these photosynthetic bacteria get the electrons they need from these alternative donors, they do not generate O_2. Commonly, these bacteria are tinted purple or deep green by their photosynthetically active pigments, an impressive sight when you encounter them in a stagnant pond (Figure 18).

FIGURE 18. Oxygen-free habitats are common on the modern Earth. Here we see a microbial community from the Turks and Caicos Islands, in the Caribbean. The dark fibrous layer at the surface (above the upper arrow), actually pigmented deep green by cyanobacteria, is exposed to the air and so is oxygen-rich. Below this veneer (in the zone between the arrows), light still penetrates, but oxygen does not, giving rise to the slightly lighter layer rich in purple photosynthetic bacteria. These bacteria use hydrogen sulfide as a source of electrons and do not generate oxygen gas. In this layer and beneath it, aerobic respiration is impossible; some microorganisms respire using sulfate and other ions instead, and others ferment organic molecules. *Andrew H. Knoll*

If photosynthetic bacteria can fix CO_2 into sugar without producing O_2, can other cells complete the carbon cycle without using oxygen in respiration? Once again, the metabolic versatility of bacteria takes center stage. You and I use O_2 to respire organic molecules, but some bacteria can respire using other compounds, such as sulfate ions (SO_4^{2-}) or oxidized iron (Fe^{3+}). That is, just as animals use oxygen generated by plants to respire organic molecules back to CO_2, these bacteria use the molecules produced when photosynthetic bacteria obtain electrons from hydrogen sulfide, dissolved iron, and the like. In this way, the carbon cycle in sunlit but oxygen-poor environments is linked to the cycles of iron and sulfur. Earth's youth may have been our first Iron Age, its carbon cycle closely tied to the biological cycling of iron in oxygen-poor rivers, lakes, and seas.

Bacteria and archaea (introduced in the last chapter as the microbial sisters to bacteria) have other metabolic tricks up their sleeves. Some use energy from chemical reactions to fix carbon, sidestepping the need for sunlight. And some gain a modest amount of energy by breaking organic molecules into simpler compounds, a process known as fermentation. You are capable of fermentation yourself, using it to generate needed energy when exercise depletes oxygen in your muscles—acid generated by this process produces the burning sensation you may feel during a strenuous workout. While you can ferment organic molecules to provide a transient source of energy, your body can't make a living this way. In fact, few cells other than bacteria and archaea are adept at fermentation, the champions being yeasts, source of the chemical magic that converts grain into beer and grapes into wine.

We see, then, that microbes alive today show us how life could have been sustained for a billion years on an oxygen-free planet. On the early Earth, diverse bacteria and archaea populated land and sea, cycling carbon, iron, sulfur, and other elements. More complicated organisms—algae, protozoans, fungi, plants, and animals—require oxygen for metabolism and so would have to wait in the evolutionary wings until O_2 became a persistent component of the Earth's surface.

WHY, THEN, DID OUR PLANET CHANGE so profoundly 2.4 billion years ago? Geologists agree on *when* O_2 began to accumulate, but at present there is no consensus on *how* it happened. Let me summarize key puzzle parts as I see them, granting that others might paint a different picture.

There is consensus on at least two points. First, oxygen in the air we breathe owes its existence to life. The only process capable of oxygenating our planet's atmosphere is oxygenic photosynthesis—photosynthesis in which water supplies electrons, generating O_2 as a by-product. Earth's Great Oxygenation Event (GOE) was revolutionary, and cyanobacteria—the only bacteria capable of oxygenic photosynthesis—were the heroes of the revolution. With this in mind, a potentially simple solution presents itself: the evolutionary origin of cyanobacteria led directly to the GOE. Simple indeed, but two observations, one geological and one ecological, suggest that the story is actually more complicated.

It turns out that sedimentary rocks older than 2.4 billion years contain chemical signatures that many interpret as evidence for transient oxygen production on a generally oxygen-free planet. Some of the same chemical signatures that record permanent environmental change 2.4 billion years ago suggest earlier accumulations of oxygen, but limited, local, and short-lived. Critics of such interpretations exist, but evidence for these "whiffs of oxygen" is abundant and growing, and if even one of them is interpreted correctly, oxygenic photosynthesis must have originated hundreds of millions of years before the GOE. Inferences from molecular biology also suggest that oxygen-producing cyanobacteria originated long before they came to dominate sunlit ecosystems.

Ecology helps to explain the geologic data. As already noted, in modern sunlit environments where dissolved iron, hydrogen sulfide, or other alternative sources of electrons are present, cyanobacteria actually don't fare well. This suggests that in early oceans, cyanobacteria were commonly at a competitive disadvantage relative to other types of photosynthetic bacteria. How could cyanobacteria have gained the upper hand in a world that long favored different photosynthetic microbes? To find an answer, we need to think beyond biology and consider the Earth itself.

This brings up the second point of consensus: the *existence* of cyanobacterial photosynthesis is not enough to foment planetary change. O_2 in the atmosphere and ocean will only accumulate when the *rate* of oxygen production by cyanobacteria exceeds the rate at which physical and biological processes remove it.

Now we have two ways of explaining how cyanobacteria could have originated long before oxygen began its permanent accumulation in the atmosphere and surface ocean. Perhaps reduced gases and ions in early oceans favored photosynthetic bacteria other than cyanobacteria. And possibly, overall rates of photosynthesis were sufficiently low so that any oxygen gas generated by early cyanos was scavenged by volcanic gases and weathering minerals. I believe that both were true.

Today, rates of photosynthesis are generally limited not by sunlight, carbon dioxide, or water, but rather by the availability of nutrients, especially phosphorus, found in DNA, membranes and ATP, the cell's energy currency, and nitrogen, required for both DNA and proteins. Some bacteria and archaea can convert nitrogen gas into biologically usable molecules, as can lightning (in limited amounts), so we'll focus on phosphorus in trying to understand the early biosphere. Phosphorus weathers from rocks exposed to the elements and enters the oceans in runoff from rivers. Photosynthetic organisms take up this phosphorus and incorporate it into biomolecules; other organisms will obtain phosphorus from the food they eat, passing it from one organism to another through the food chain. Eventually, much of this phosphorus sinks to the seafloor, in a slow rain of organic particles from the surface. Bacteria within the sediments free up much of this phosphorus, and deep sea currents return it to the surface to fuel renewed photosynthesis.

In early oceans, phosphorus input from the continents was low because of the limited volume of rocks sticking out above the sea. At the same time, the return of phosphorus to the surface by the upwelling of deep waters would also have been limited by inefficient recycling. My lab and others have used basic principles of chemistry to estimate how much phosphorus would have been available to photosynthetic microorganisms in early oceans, and the answer is "not much." In fact, nutrient availability probably placed strong constraints on early life, limiting photosynthesis, whether by cyanobacteria or other bacteria, to levels unlikely to affect global transformation.

As our planet matured, large, stable continents emerged

above sea level, increasing the erosional flux of phosphorus to the oceans. Eventually, as the phosphorus supply came to outstrip the availability of alternative electron donors, cyanobacteria gained ecological importance. And as they did so, they transformed the world. The oxygen they produced scavenged other sources of electrons from sunlit waters, permanently tipping the biosphere toward oxygenic photosynthesis and oxygen-rich air. Now, as sediments buried organic matter produced by cyanobacteria, shielding it from respiration, Earth's engine for O_2 accumulation was engaged. There was no going back.

In this view, the Great Oxygenation Event was not simply a product of Earth's physical development; nor did it exclusively reflect evolutionary innovation. It was the *interaction* between Earth and life that transformed our planet's surface.

HOW MUCH OXYGEN ACCUMULATED during the GOE and its aftermath? And what were its consequences? Quantification of ancient oxygen levels remains challenging, but several observations indicate that the answer, once again, is "not much." Chemical analyses of sedimentary rocks indicate that for nearly two billion years following the GOE, the world's oceans looked something like the modern Black Sea, with oxygen in surface waters, but not below. While some data suggest that oxygen rose markedly during the GOE, by about 1.8 billion years ago, O_2 in the atmosphere and surface ocean had settled back to perhaps 1 percent or so of modern values—plenty to support an amoeba, but not enough to sustain a beetle. (Iron formation returned briefly but globally about 1.9 billion years ago, perhaps reflecting a strong pulse of hydrothermal fluids from the mantle into the oceans. Iron mined in the Mesabi Range of Minnesota reflects this event.)

Even at modest concentrations, however, oxygen offered new possibilities for life. Fueled by cyanobacteria, ecosystems grew more productive and more energetic. (Respiration using O_2 yields much more energy than respiration without oxygen or fermentation.) And if you could be transported back to this brave new world of oxygen gas, microscope in hand and protected by face mask and oxygen tank, you'd have noticed something that wasn't there earlier. Halfway through the history of life, a new type of cell emerged.

Eukaryotes are organisms whose DNA is compartmentalized within a nucleus. You are a eukaryote, as are ponderosa pines, seaweeds, mushrooms, and singled-celled organisms that range from amoebas to diatoms—perhaps ten million species altogether. While the nucleus defines eukaryotes, other cell features play key roles in their history and ecology. Particularly important, eukaryotes, unlike bacteria, have a dynamic internal system of molecular scaffolding and membranes that allows their cells to grow large and take many different shapes. It also enables eukaryotes to make a living in ways that bacteria generally can't, in particular by engulfing small food particles, including other cells. Through predation, then, eukaryotic cells brought new complexity to ecosystems. And, as we'll discuss in the next chapter, new ways of communicating among cells paved the way for complex multicellular organisms.

Within eukaryotic cells, respiration and photosynthesis are localized within small structures called organelles; mitochondria are the seat of respiration, chloroplasts the site of photosynthesis. These organelles look a bit like bacterial cells. Chloroplasts, for example, have internal membranes much like those of cyanobacteria. More than a century ago, the Russian botanist Konstantin Mereschkowski argued that this similarity was no coincidence. Aware of the earlier discovery that reef corals harbor algae within their tissues, Mereschkowski argued that chloroplasts originated as once free-living cyanobacteria that were engulfed by protozoans and eventually reduced to metabolic slavery. When not ridiculed, Mereschkowski's idea was simply forgotten—a common fate in science. In this case, however, Mereschkowski was right. As the age of molecular biology dawned, it became possible to revisit his hypothesis with new tools. The chloroplast contains a small amount of DNA, and molecular sequence analysis of its genes makes it clear that, in the Tree of Life, chloroplasts nest within the cyanobacteria. Further research showed that mitochondria, too, have bacterial ancestry. It increasingly looks like the eukaryotic cell itself emerged from a long-ago partnership between an archaean cell and a bacterium capable of aerobic respiration. Indeed, scientists have recently discovered archaea that contain molecules similar to those that organize the cell interior in eukaryotes. We are evolutionary chimeras, and plants have an additional partner, harnessing the power of cyanobacteria to bring photosynthesis to our domain.

Let's put this biological story into environmental perspective. Most eukaryotes respire using oxygen, and those that don't are descended from ancestors that did. Moreover, nearly all eukaryotes that live where oxygen is absent still require biomolecules that form only where O_2 is available; they get what they need by eating food sourced from oxygen-rich habitats. So, in an important way, eukaryotes are daughters of the GOE.

Consistent with this view, we begin to see fossils of eukaryotic cells in sedimentary rocks deposited 1.6–1.8 billion years ago. Rocks of this age from Australia, China, Montana, and Siberia all contain a modest diversity of microfossils whose preserved cell walls exhibit a structural and morphological complexity found today only in eukaryotic organisms. Some had long arm-like extensions, perhaps enabling them to absorb dissolved organic molecules, much as fungi do today (Figure 19). Others had thick plate-like walls, enabling them to lie dormant when the environment didn't favor growth (Figure 20). A few even attained a simple degree of multicellularity, forming sheets of cells visible to the naked eye (Figure 21). A new biological revolution was afoot, but we should remember that emerging eukaryotes did not replace the bacteria and archaea that had ruled the Earth since life began. Eukaryotes were intercalated into microbial ecosystems still dependent on microbial metabolism. Even today, the biosphere has 30 tons of bacteria and archaea for every ton of animal.

Climbing upward through the next billion years of preserved fossils, we find more and more eukaryotic diversity, including unambiguous algae descended from that early partnership between a protozoan and a cyanobacterium, cells protected from predators by tough vase-shaped walls or scaly armor, and a growing diversity of simple multicellular structures (Figures 22 and 23).

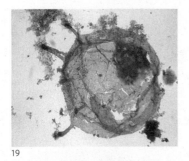

19

20

50 μm

21

FIGURES 19–21. Fossils of early eukaryotic organisms. (Figure 19) a single-celled organism with arm-like extensions that may have functioned to absorb organic molecules for food, from 1,500- to 1,400-million-year-old rocks in northern Australia; (Figure 20) a thick, plate-like cell wall that would have protected its owner from an unfavorable environment and other organisms, also from 1,500- to 1,400-million-year-old rocks in Australia; (Figure 21) among the oldest known organisms with a simple multicellular structure, from nearly 1.6 billion-year-old rocks in China. The bar in 20 = 50 microns in 19 and 20, and = 5 millimeters in 21. *Figures 19 and 20 by Andrew H. Knoll; Figure 21 courtesy of Maoyan Zhu, Nanjing Institute of Geology and Palaeontology*

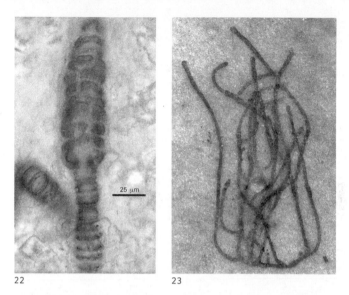

22   23

FIGURES 22 AND 23. Fossils show that diverse eukaryotes thrived before the dawn of animals. Here we see the oldest known red (Figure 22) and green (Figure 23) algae, preserved in billion-year-old rocks from arctic Canada and China, respectively. Bar in Figure 22 = 25 microns for that figure, and = 225 microns for Figure 23. *Figure 22 courtesy of Nicholas Butterfield, University of Cambridge; Figure 23 courtesy of Shuhai Xiao, Virginia Tech*

This world of low oxygen and (mostly) microbial life persisted for many millions of years, but among those simple multicellular creatures in late Proterozoic oceans, another revolution was brewing. In uppermost Proterozoic rocks, deposited on the heels of a vast global ice age, large complex organisms appear in the fossil record. More than three billion years after life emerged, the age of animals was at hand.

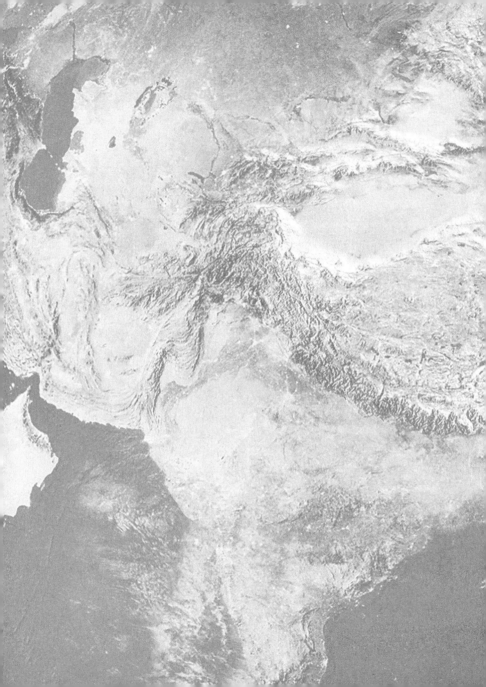

5

Animal Earth

LIFE GETS BIG

HAPPY THE PALEONTOLOGIST who visits Mistaken Point on a sunny afternoon. A UNESCO World Heritage Site along the rocky coast of southeastern Newfoundland, Mistaken Point is commonly shrouded by fog or pelted by a driving rain. However, if you arrive late on a rare clear afternoon, when low-angle sunlight throws the surface features of ancient beds into sharp relief, you'll never forget the sight.

The cliffs at Mistaken Point consist of muddy sediments and volcanic ash laid down, one bed after another, on the deep seafloor, about 565 million years ago. The site has three remarkable features that collectively make it special. First, step-like sea cliffs expose expansive surfaces of ancient sedimentary beds, preserved for all time by rapid burial and essentially enabling us to walk across the ancient seafloor. The second unusual feature is the abundance of volcanic ash, which facilitates the dating of individual beds. Third—and most extraordinary—is what populates the bedding surfaces. Once you get your eye in, you'll see weird and wonderful fossils by the hundreds, seemingly alien life forms preserved where they lived, entombed by volcanic ash—a paleontological Pompeii (Figure 24). Some look like

FIGURE 24. Fossils of early animals in 565-million-year-old sedimentary rocks from Mistaken Point, Newfoundland. Scale bar shows 1 centimeter units. *Courtesy of Guy Narbonne, Queen's University*

fronds, others are fan-like. A few are long and thin, a bit like the tail feathers of pheasants. Many stood erect above the seafloor, anchored to the sediment by a bulbous holdfast, but swaying in the current. Others spread out across the sediment surface. But whatever their length and width, all are only a few millimeters thick, and most have a quilted structure, a bit like the conjoined tubes of the air mattress I took camping as a boy.

Perhaps surprisingly, most scientists accept these as the oldest known fossils of animals, our first paleontological glimpse of the group that would diversify across the entire face of the planet.

To understand the biology and, hence, evolutionary relationships of Mistaken Point fossils, we need to start from first principles, paying close attention to what is preserved—and, equally, what is not. Let's begin by examining how these strange organisms gained carbon and energy. How did they make their living? Some look superficially like seaweeds, so maybe they were photosynthetic. No, the Mistaken Point organisms lived several hundred meters beneath the sea surface, well below the depths penetrated by sunlight. Today, some deep sea animals harness the power of symbiotic bacteria that can use chemical energy to fix carbon. But, this won't work, either, as animals that live in close association with such bacteria thrive where oxygenated and oxygen-free waters meet. Chemical evidence from Mistaken Point rocks indicates that these organisms lived in stable, relatively oxygen-rich environments.

What's left is heterotrophy—gaining carbon and energy by eating organic molecules originally synthesized by other species. We're heterotrophs, as are sharks, crabs, and squids. But this list draws attention to features the Mistaken Point fossils lack. They don't have mouths, and they don't have limbs to move around or grab prey. They don't seem to have had a well-developed digestive system, and few if any of them moved actively on or above the seafloor. How, then, could they feed?

At this point, we need to return to living animals for points of comparison, though not to the species we encounter every day in forests, zoos, or nature films. Allow me to introduce *Trichoplax adhaerens*, the only formally described species of the obscure phylum Placozoa (Figure 25). Among the world's smallest (a few millimeters long) and simplest animals, *Trichoplax* individuals consist largely of upper and lower cell sheets, called epithelia, that sandwich an interior containing fluid and a few fibrous cells; they have no mouth, no limbs, no lungs, gills, kidneys, or digestive system. Cells that line *Trichoplax*'s surface can engulf food particles, much the way protozoans do, and they also absorb organic molecules from surrounding water or sediments. They get the oxygen they need by diffusion, and for this reason are constrained to be thin.

This thumbnail description of *Trichoplax* sounds more than a little like our earlier portrait of the Mistaken Point fossils, save for size. In fact, I subscribe to a view introduced in 2010 by then graduate students Erik Sperling and Jakob Vinther that living placozoans may be the lone survivors of the early

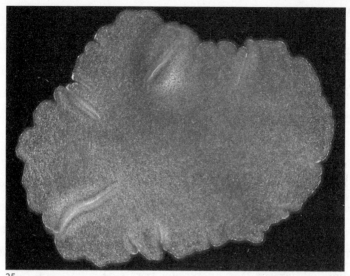

25

FIGURES 25 AND 26. *Trichoplax adhaerens* and its proposed evolutionary relationship to both Ediacaran and living animals. *Figure 25 courtesy of Mansi Srivastava, Harvard University*

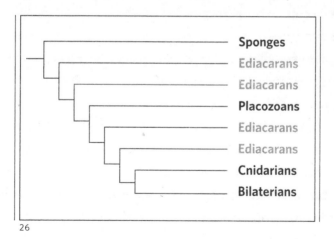

Sponges
Ediacarans
Ediacarans
Placozoans
Ediacarans
Ediacarans
Cnidarians
Bilaterians

26

radiation of animals documented by fossils at Mistaken Point and elsewhere. Figure 26 shows a simple phylogeny—a family tree—of animals. Focusing for the moment on extant groups, the tree indicates that the last common ancestor of these animals gave rise to two lineages, one containing sponges and a second that includes pretty much everything else. A few Mistaken Point fossils suggest affinities with sponges, but these were not ecologically prominent in the local ecosystem. Climbing the "everything else" limb of the tree, we come to a second branch point from which placozoans diverge and then another where cnidarians (sea anemones, corals, jellyfish) and bilaterians (insects, snails, us, and everything else with a head and tail, top and bottom, left and right) part ways. The logic of this tree requires that branch points closer to the base of the tree predate those higher in its crown. Given the comparison of our Newfoundland fossils to placozoans, this suggests that the peculiar fossils of Mistaken Point—more Dalí than Devonian—reflect an early diversification of anatomically simple animals that postdated the divergence of sponges but preceded expansion of the more complex and diverse cnidarians and bilaterians so conspicuous in modern oceans.

Mistaken Point remains are sometimes called Ediacaran fossils—or just Ediacarans—because they lived during the

Ediacaran Period. The Ediacaran Period, established as part of the international geologic timescale only in 2004, more than a century after the periods of the ensuing Phanerozoic Eon, is bounded by two events of enormous import. Beginning some 80 million years before the Ediacaran Period commenced, Earth was twice enveloped from pole to equator by glacial ice, forming a "Snowball Earth." Very likely, Earth's greatest ice ages had a profound effect on biology, and indeed, many algae and protozoans recorded as fossils in pre-glacial oceans do not occur in rocks deposited after the ice retreated. But many lineages must have survived, including the ancestors of Ediacaran (and all living) animals. How did Earth get into this deep freeze and, as important, how did it get out again?

Geologists and climate modelers continue to debate the causes of late Proterozoic ice ages, but all agree that, as ever, the carbon cycle played a decisive role in the extreme climatic events recorded by the rocks. One attractive hypothesis for Snowball onset points to a massive outpouring of volcanic rocks across low-latitude continents. Volcanic rocks consume a great deal of CO_2 as they weather, and the warm temperatures found near the equator would ensure rapid weathering and erosion. Thus, tectonic events may have reduced the greenhouse gas carbon dioxide to low levels, cooling the planet and so initiating glaciation. In 1969, the Russian climatologist Mikhail Budyko postulated that as ice spreads from the poles toward the equator, it will reflect more of the sun's incoming radiation back into space, cooling the planet and, hence, facilitating further expansion of ice sheets (and still more reflection of sunlight back into space). In time, runaway glaciation should envelop the Earth. Budyko argued that, despite its mathematical plausibility, this never happened, because, he reasoned, once Earth entered a Snowball state, it could never escape. Geology indicates that just before the Ediacaran Period the entire Earth turned white. Thick ice sheets spread across the continents from pole to equator, and sea ice blanketed the oceans—think of an Antarctic landscape spreading across the Caribbean. But, you are living proof that Earth did escape these icy clutches.

Evidence from the rocks shows that after millions of years, the ice disappeared quickly, as glaciers retreated to the poles and mountaintops, and then disappeared. What precipitated the collapse of these great ice sheets? Once again, we turn to the carbon cycle. As ice spread across the planet, processes that remove carbon dioxide from the atmosphere—mainly continental weathering and photosynthesis—slowed to a trickle, but processes that add CO_2 to air—mostly volcanism—continued apace. Carbon dioxide in the atmosphere built up through time, eventually reaching the critical level at which greenhouse warming set off catastrophic melting. Out with the ice age, in with the Ediacaran Period.

MISTAKEN POINT HOLDS pride of place as one of the oldest known occurrences of Ediacaran animals, but several dozen other localities from as far afield as Australia (home of the Ediacara Hills, which lend their name to the period), Russia, China, northwestern Canada, California, and Africa show that animals broadly similar to those from Newfoundland flourished globally in late Ediacaran oceans. There's *Dickinsonia*, flat ovals that hugged the seafloor 560–550 million years ago (Figure 27). There are obvious differences from Mistaken Point fossils, but the general view holds: these were simple organisms made up of repeated, probably fluid-filled tubes that fed by particle capture and absorption and gained oxygen by diffusion. Interestingly, some exceptional specimens from the White Sea region of Russia preserve molecular fossils that confirm *Dickinsonia*'s place among the animals.

And then there's *Arborea,* a frond-like fossil found widely in younger Ediacaran sandstones (Figure 28). The *Arborea* animal had a circular holdfast tethered to the shallow seafloor and a

cylindrical stalk that raised two feather-like flanges into surrounding water. With no obvious mouth, gills, digestive system,
or limbs, it is probable that *Arborea* also fed and obtained oxygen much like *Dickinsonia* and Mistaken Point animals. But one
feature sets *Arborea* apart. Careful study by Frankie Dunn and
her colleagues shows that each bulbous unit on its flanges connects to a thin tube that descends through the stalk to its base.
This and the generally modular construction of the fossil open
the possibility that *Arborea* was a colony rather than a single
individual. That's not surprising, as, before the evolution of bilaterian animals with well-developed organs, colony formation
would have been nature's foremost way of generating animal
complexity. Take, for example, the Portuguese man-of-war, a
present day sea-going cnidarian whose stingers leave ugly welts
on careless swimmers. The man-of-war looks like a jellyfish but
in fact is a colony made of numerous individuals, each of which
has developed in a way that supports a specific function. The
conspicuous float is one individual. Tube-like structures that
hang downward from the float are separate individuals dedi-
cated to feeding, reproduction, or defense. *Arborea* may record
an early evolutionary experiment in this direction.

Not all late Ediacaran fossils fit this mold, however. *Kimberella* is a small fossil first discovered in Australia but best

known from more than a thousand specimens, beautifully preserved in White Sea rocks (Figure 30). A few centimeters long,
Kimberella had a distinct front and back, top and bottom, left
and right, identifying it as part of the great bilaterian limb on
the animal tree. Its fossils reveal evidence of a muscular foot,
with overlying viscera and a lightly ornamented covering. Ancient trackways make it clear that Kimberella moved around on
the seafloor, while scratch marks that radiate from its mouth
area tell us that these animals had a tough, comb-like organ in
their mouths that enabled them to feed by scraping algae and
other microorganisms from the seafloor, much like the radula
of living snails. Tracks and trails in other sandstones of this
age record additional simple bilaterians, known only from their
movements on and within the seafloor (Figure 31).

Innovations continued as the Ediacaran Period approached
its end. Tubes of calcium carbonate, first discovered in 547- to
541-million-year-old limestones from Namibia but, again, known
to occur globally, mark the emergence of mineralized skeletons
in animals (Figure 29). Such armor costs energy to make, but as
predators expanded, the expense of construction was rewarded
by the priceless dividend of survival. By the end of the period,
Ediacaran oddities were moderately diverse, but the familiar
animals of our own world were yet to come.

27

28

29

30

31

As animals radiated into Ediacaran oceans, the world around them was changing as well, laying the foundation for our modern biosphere. We've already noted that through most of the Proterozoic Eon, O_2 levels in the atmosphere and surface oceans were low, perhaps as little as 1 percent or so of today's levels. Comparably oxygen-poor environments persist today in a few corners of the ocean; they host animals, but mostly tiny (up to a few hundred microns long and a few tens of microns wide) species unlikely to enter the fossil record. Large, diverse, and energetic animals—including carnivores, about to take center stage in our narrative—occur only where oxygen levels are higher. Macroscopic animal fossils, then, suggest that during the Ediacaran Period, our planet underwent a profound environmental sea change (literally!), and thousands of chemical analyses by dozens of labs provide independent evidence that, at this time, Earth began its protracted transition to the oxygen-rich planet we inhabit today.

FIGURES 27-31. Fossils and trackways of animals in Ediacaran rocks, including *Dickinsonia* (Figure 27), *Arborea* (Figure 28), the earliest mineralized animal skeleton (Figure 29), *Kimberella* (Figure 30), and tracks made by an early bilaterian animal with limbs (Figure 31). *Figure 27 courtesy of Alex Liu, University of Cambridge; Figure 28 courtesy of Frankie Dunn, University of Oxford; Figures 29 and 31 courtesy of Shuhai Xiao, Virginia Tech; Figure 30 courtesy of Mikhail Fedonkin, Geological Institute, Russian Academy of Sciences*

As animals grew large and oxygen became plentiful, changes were also afoot among Earth's photosynthetic biota. Both fossils and preserved lipids indicate that after more than three billion years of largely bacterial photosynthesis, algae rose to ecological prominence in the oceans. How might we explain this coordinated transition in animals, algae, and air? There is reason to believe that, driven by large-scale Ediacaran mountain building, more nutrients became available in the oceans. In the modern sea, cyanobacteria continue to be important members of the plankton where nutrients are scarce, but eukaryotic algae tend to dominate where nutrient levels are higher. The pattern we see today in space suggests what happened during Ediacaran time. More nutrients, more photosynthesis by diversifying algae. More photosynthesis, more food and oxygen, and—more than three billion years after life began—a world capable of supporting large, energetic animals.

IF ICE BOUNDED the Ediacaran Period from below, it is evolution that frames it from above. To see evidence of this, we need to travel about 2,800 miles west of Newfoundland to the small town of Field, British Columbia, just west of Canada's all-world scenic treasure, Lake Louise. On a mountainside high above the valley floor, paleontologists carefully prise slabs of dark shale from a small quarry. More often than not, their labors are rewarded by shiny compressions of animals (and a few algae), preserved in remarkable anatomical detail on slab surfaces. The rocks, known as the Burgess Shale, were deposited 510–505 million years ago as mud, stirred by storms or jostled by earthquakes, sped down a steep slope to accumulate on the relatively deep seafloor. The mud buried myriad organisms, sealing them from decay by hungry microbes. As a result, we see not only the mineralized skeletons that comprise conventional fossil deposits but unmineralized carapaces, limbs, gills, digestive tracts, and even nerve ganglia, laid out like pictures in an ancient anatomy text.

And what organisms they were (Figures 32–34). The Cambrian Period (541–485 million years ago) is famous as the interval during which we first see abundant fossils of familiar-looking animals. The conventional record of Cambrian animals, recorded by mineralized shells and other skeletons, is dominated by extinct arthropods called trilobites; these segmented, multilimbed creatures comprise some 75 percent of all fossil species discovered in Cambrian rocks. At Burgess, trilobite fossils are also abundant (Figure 32), but arthropods as a whole make up only a third of Burgess species diversity, and most

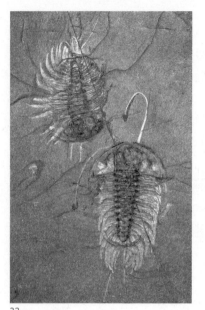

FIGURES 32-34. Cambrian fossils from the Burgess Shale. Trilobites, showing exquisitely preserved limbs and antennae (Figure 32); *Opabinia,* an extinct relative of arthropods (Figure 33); and a polychaete worm with conspicuous bristles (Figure 34). *Copyright Smithsonian Institution—National Museum of Natural History. Photographs by Jean-Bernard Caron.*

32

of these ancient arthropods are not trilobites but rather wonderfully weird forms that did not precipitate minerals in their exoskeletons and so do not preserve under most conditions. Sponges are common, and trained biological eyes can spot representatives of numerous bilaterian phyla, including mollusks (snails, clams, squid), polychaete and priapulid worms, even close cousins of our own group, vertebrate animals. Other formations from China, Greenland, and Australia expand this remarkable window on our biological past and deepen it in time to at least 520 million years ago.

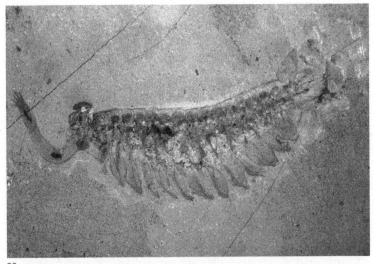

33

34

Ediacaran and Cambrian fossil assemblages are strikingly different, but could it be that the observed biological distinctions reflect biases of preservation and environment rather than evolution? It turns out that we can reject this possibility. First off, in circa 550-million-year-old shales from China, Burgess-style preservation documents diverse macroscopic Ediacaran organisms. There are lots of seaweeds and a few possible animals, but no trace of the arthropods, mollusks, and other complex bilaterians to come. Trace fossils tell a similar story. Mobile animals leave a calling card in the form of tracks, trails, and burrows that reflects both anatomy and behavior. A limited diversity of simple traces can be found in later Ediacaran rocks, but, again, nothing like the complex tracks and burrows that mark Cambrian sandstones and shales. And while mineralized skeletons occur in late Ediacaran rocks, their simple morphologies and limited diversity pale before the rich record of Cambrian skeletons.

Clearly, then, biological differences between the Mistaken Point and Burgess biotas reflect a remarkable interval of animal diversification, commonly called the Cambrian Explosion. Without question, Cambrian fossils record the emergence of a new biosphere that is both the culmination of and a major departure from the preceding three billion years of evolution.

If we look carefully at Cambrian fossils we begin to discern differences from living animals as well as similarities. In his best-selling book *Wonderful Life*, the late Stephen Jay Gould focused on the differences, seeing Burgess animals as "weird wonders" that record extinct body plans. His favorite example was *Opabinia*, a little creature about 2–3 inches (4–7 centimeters) long, with five eyes and a long, flexible proboscis that ends in a claw (Figure 33). Weird? Absolutely. But alien? Probably not. Despite its curious details, *Opabinia* had a segmented body with a tough organic exoskeleton, much like those of arthropods. Other fossils in Cambrian rocks also show combinations of the strange and the familiar, and when we put them all together, they show how the body organization we recognize as arthropod came to be. Viewed through the lens of Cambrian fossils, then, living arthropods can be thought of as the (very successful!) survivors of a broader Cambrian lineage. And what is true for arthropods holds true for other phyla, as well. Cambrian fossils provide snapshots of animal body plans as they took shape.

The Cambrian Period, then, stands out as transitional. The great leap forward of the Ediacaran Period continued and, indeed, accelerated during the Cambrian Period, but it didn't leave us with a fully modern biosphere. Fossils document diverse animal body plans *in statu nascendi* but with few species and not many fully modern forms. A number of animal groups evolved skeletons hardened by minerals, armoring their bodies against rapidly diversifying carnivores, but Cambrian limestones still formed mostly by physical or microbially facilitated calcium carbonate precipitation. (Today, skeletons account for most limestone deposition in the oceans.) Reefs that dotted the shallow seafloor were built mainly by microbes, although fossils make it clear that animals thrived in and around these structures. Seaweeds were relatively common, but, like their animal counterparts, algal fossils display limited diversity. Oxygen in air and oceans was richer than it had been in the past but still no more than about half of modern levels; deep ocean waters remained oxygen-free. Several lines of evidence tell us that Cambrian climates were warmer than today's, a real greenhouse after the protracted ice house climate of the Snowball Earth. Swimming in a Cambrian ocean, you would have been dazzled by the many animals that darted through the water column, intent on catching prey or avoiding capture. At the same time, however, you might be perplexed by the thorough mixture of alien and ordinary, both among species and within individuals. I'm reminded of bas reliefs in ancient Egyptian temples—it is tempting to interpret them through modern eyes, but probably unwise.

YEARS AGO, I had the illuminating experience of hiking through a thick stratigraphic section of limestones that record life and environments during the Ordovician Period (485–444 million years ago), which followed the Cambrian. Rocks near the bottom of the succession look much like those from the underlying Cambrian—relatively few fossils, with a limited diversity of most groups other than trilobites. But, as I climbed upward, ascending through the Ordovician record, the rocks slowly began to change. There were still plenty of trilobites, but now also other skeletal fossils in abundance.

To catch a glimpse of this emerging world, take a drive along the country roads around Richmond, Indiana, a small town northwest of Cincinnati best known as the home of Earlham College. Bulldozed cuts along the roadsides expose late Ordovician (ca. 450–445 million years old) limestones and shales that drip with fossils. And these fossils no longer have an alien aspect; they are recognizably the skeletal remains of clams, snails, cephalopods (the group to which squids and the octopus belong), corals, bryozoans (moss animals), brachiopods, and sea lilies. Locally, these skeletons built upward from the sea-floor, forming patch reefs at least broadly similar to those you can see by snorkeling along the Florida Keys or in the Bahamas. Rocks of this age display no new phylum-level body plans, but species diversity increased dramatically—by nearly an order of magnitude, according to some estimates. And for the first time, skeletons emerged as major components of limestones formed on the shallow seafloor.

Explanations for this third stage of marine animal diversification are many and varied. Some geologists point to chemical evidence for cooling oceans, potentially enhancing ecological opportunities for animals. Others argue that oxygen levels increased, providing another physical spur to animal diversification. Still others postulate ecological drivers, suggesting that increased predation pressure underpins the diversification of heavily skeletonized animals and algae.

All of these explanations may be true, but individually each is probably incomplete. Once again, the physical and biological processes at work in the biosphere did not act independently of one another. The evidence for global cooling is robust, perhaps driven by rising mountains whose weathering quickened the removal of CO_2 from the atmosphere. Cool waters can accommodate more dissolved oxygen than warm ones, and so Ordovician cooling would result in greater O_2 availability for animals in the shallow oceans, even if the atmosphere remained unchanged. And carnivorous animals generally require more oxygen than other kinds of animals, because predation takes a lot of energy.

Whatever the correct interpretation, Late Ordovician oceans teemed with animals. Extinct corals, massive bryozoans (unlike any seen today), and heavily mineralized sponges constructed reefs, which provided food and shelter for diverse predators and scavengers, including conical relatives of squids—some as much as 3.5 meters (11 feet) long—and fish, recognizable by their fins and tails but without jaws. Global cooling culminated in a brief but substantial ice age, recorded by glaciogenic rocks in what are now Southern Hemisphere continents. And something else happened. By the time the glacier collapsed, some 70 percent of all known animal species had disappeared.

Green **Earth**

PLANTS AND ANIMALS COLONIZE LAND

IN 1991, I boarded an aging Aeroflot jetliner in Moscow, bound for Yakutsk, a Siberian city 3,000 miles (4,900 kilometers) to the east. For much of the eight-hour flight, I peered out the window, seeing little below but a seemingly endless expanse of forest, broken only by the silver threads of rivers as they meandered toward the Arctic Ocean. During the Cambrian Period, as trilobites reveled in their evolutionary youth, a similar flight would have traversed mostly bare rock, here and there tinted by microbial slime. The green of Siberian landscapes, then, reflects another biological revolution, the colonization of land by complex multicellular organisms.

Microbes probably took root on land early in Earth history, but it is plants that changed the world, providing both food and physical structure for complex terrestrial ecosystems. Today, some 400,000 species of land plant account for half of Earth's photosynthesis and an estimated 80 percent of our planet's total biomass. Indeed, Earth's resplendent robe of green is such a pervasive feature of our planet that it can be detected from space. In 1990, as NASA's Galileo spacecraft winged toward Jupiter, it trained its mechanical eyes on the distant Earth, revealing in our planet's reflected light a distinctive peak in the near-infrared—the so-called Vegetation Red Edge. This signature arises because land vegetation strongly absorbs incoming visible radiation but reflects infrared wavelengths back to space. Visitors to the early Earth would have observed no such feature.

Animals, although born in ancient oceans, are today most diverse on land—insect species alone far outnumber all animal species in the sea. A huge and mostly undocumented diversity of fungi permeates soils, and myriad protists and bacteria cycle carbon, nitrogen, sulfur, and other elements on land, much as they have long done in water.

Clearly, our familiar world of fields and forests, grasshoppers and rabbits reflects a remarkable transformation of continents and islands, carried out only in the most recent 10 percent of Earth's history. How did the greening of our planet take place, and what were its consequences for Earth itself?

IN 1912, WILLIAM MACKIE, a physician by training, passed through the village of Rhynie, Scotland, while surveying the regional geology. Rhynie, about thirty miles northwest of Aberdeen, sits among rolling fields, with few rocky outcrops to divert the geologist, so when Mackie noticed some unusual stones in the walls bounding local fields, he stopped to take a closer look. The rocks were made of chert (SiO_2), and even cursory examination showed them to contain what appeared to be fossil stems, some preserved in growth position. Mackie had discovered the Rhynie Chert, paleobotany's counterpart to the Burgess Shale. Deposited 407 million years ago in and around hot springs similar to those found today in Yellowstone or the North Island of New Zealand, Rhynie provides a remarkably clear glimpse of terrestrial ecosystems in evolutionary adolescence.

Plants are the acknowledged stars of Rhynie's show, although, as we shall see, they share the stage with myriad other kinds of organisms. Many features of cell and molecular biology make it clear that land plants evolved from green algae that lived in fresh water. But the evolutionary journey from rivers and ponds to dry land involved real challenges, including the need to avoid desiccation, mechanical support, and resource acquisition. When surrounded by water, photosynthetic organisms are not in danger of drying out, but on land, water vapor constantly evaporates from cells—put the fresh water cousins of plants on dry land and they'll quickly shrivel and die. Photosynthetic life on land thus required a means of retarding evaporation from living tissues. Aquatic algae need no special tissues to remain vertical on a lake or river bottom because water itself buttresses their bodies. But on land, air can't support erect tissues, so plants need other ways of maintaining their stature. And in lakes and rivers, nutrients are absorbed from surrounding water, whereas on land, they must be acquired from soil and transported to sites of cell growth. Fortunately for paleontologists, the adaptations evolved by plants for life on land are largely anatomical—you can see them in living plants, and they preserve well in fossils.

Rhynia, an iconic early plant that covered much of the Rhynie landscape, consisted mainly of naked photosynthetic axes, pencil-size structures that grew along the ground, much like the runners of strawberries, and episodically produced vertical branches that reached up to eight inches (20 centimeters) into the sky (Figure 35). The axes preserve a thin external coating of waxes and fatty acids called cuticle. As known from living plants, cuticle effectively prevents water vapor in cells from escaping into the atmosphere, but it does an equally good job of keeping carbon dioxide from diffusing into the plant for photosynthesis. And so, like living plants, *Rhynia* displays an elegant solution to the problem of balancing CO_2 gain against water loss. Its surface has numerous small holes called stomata, flanked by cells that expand to seal the opening when the plant suffers water stress and then contract to reopen the holes when it is safe to do, allowing carbon dioxide into the plant. Cuticle with stomata, thus, is a sine qua non of land plants, preserved for all to see in Rhynie fossils.

Because photosynthesis on land inevitably entails water loss, plants need a mechanism of absorbing water from their surroundings and transporting it through the plant. In terrestrial ecosystems, water, along with nutrients such as nitrogen and phosphorus, resides mainly in the soil. Living plants develop roots that spread their thin, finger-like projections through the substrate, taking up both water and nutrients. Actually, for most plants, much of the work of nutrient uptake is carried out by fungi that live in close association with roots. Rhynie plants didn't have well-developed roots, relying instead on thin filaments called rhizoids to anchor them to the ground and absorb water. But the fossils show that more than 400 million years ago, land plants already lived in close association with fungi, exchanging food for nutrients. In the absence of this partnership, Earth's green revolution might never have occurred.

Finally, plants need to transport water and nutrients upward from the ground and move food generated by photosynthesis throughout the plant body, feats accomplished by specialized tissues called the vascular system. At the same time, water-conducting cells have thick walls that provide mechanical

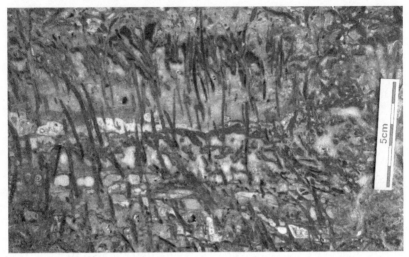

35

FIGURES 35-37. The Rhynie Chert, 407 million years old, Scotland. Rhynie rocks provide one of our earliest glimpses of terrestrial ecosystems, including simple plants (Figure 35, anatomically preserved cross section in Figure 36), animals, fungi (Figure 37, arrows point to fungi living on the tissues of Rhynie plants), algae, protozoans, and bacteria, all living on land or in shallow pools. *Figure 35 courtesy of Alex Brasier, University of Aberdeen; Figure 36 courtesy of Hans Steur; Figure 37 courtesy of Paleobotany Group, University of Münster*

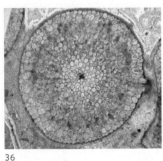

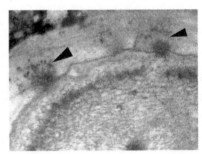

36   37

strength for the plant axis. Anatomical sections of *Rhynia* reveal a thin cylinder of vascular tissue that ran upward through the center of the plant's axis (Figure 36).

Erect *Rhynia* axes commonly terminate in an elongated compartment that contains spores for reproduction. In water, spores can swim from one place to another, making dispersal a relatively simple matter. On land, however, *Rhynia* spores were blown across the landscape by wind, exposing them to desiccation. Like modern fern spores or pollen grains, *Rhynia*'s spores were coated by a complex polymer called sporopollenin, which inhibits water loss while providing "sunglasses" against harmful UV radiation. In overall anatomy, then, *Rhynia* and other Rhynie plants resembled living plants, but from there things get more interesting, as they lacked leaves, large roots, wood, or seeds. All in all, much as Burgess does for animals, the Rhynie Chert preserves photosynthetic pioneers caught in the act of becoming plants.

More than a dozen animal species have also been discovered at Rhynie, all arthropods save for one remarkable occurrence of nematodes, tiny roundworms that are simultaneously among Earth's most numerous animals and its rarest fossils. Like early plants, animal colonists had to avoid desiccation and support their weight on land. A waxy coating on the organic exoskeletons of arthropods conserved water, while their muscular jointed legs, originally evolved in the sea, provided a sturdy means of locomotion and structural support on land. Oxygen was another challenge, as gills work fine in water but aren't much use in air. Many scorpions and spiders breathe using book lungs, intricately folded tissues that maximize surface contact with air, allowing oxygen to diffuse from the atmosphere into blood-like fluids that carry O_2 throughout the body. Book lungs appear to have been derived from gills in aquatic ancestors.

Rhynie rocks additionally contain the oldest known insects, a starting bell for a radiation that would come to dominate the animal kingdom in terms of diversity. They also preserve diverse fungi, both those that ate dead plants and those that sustained live ones (Figure 37). And there are oomycetes, fungus-like microorganisms best known as the cause of potato blight in nineteenth-century Ireland; amoebae that built vase-like organic tests around their cells; green algae and cyanobacteria. In short, Rhynie fossils show by 400 million years ago, terrestrial ecosystems had already established the rudiments of their modern ecological structure and diversity. Fragments of older fossils show that early precursors of land plants established a terrestrial foothold some 50 million years before Rhynie, and, moving forward in time, over the 50 million years after our

Scottish milepost, plants underwent a remarkable evolutionary burst, evolving leaves, roots, wood, and seeds, all but the last in several distinct lineages.

Vertebrates, our own ancestors, were relatively late comers to the party. Land vertebrates, called tetrapods because they have four limbs, are clearly descended from fish that first diversified as part of the Cambrian explosion in the oceans. Indeed, comparative biology and molecular sequence analyses show that tetrapod vertebrates are close relatives of a specific group called the lobe-finned fishes. Most bony fish, from tuna to trout, have fins supported by long, thin bones that fan out from a set of small bones attached to the body. In contrast, lobe-finned fish have paired fleshy fins attached to the body by a single bone, with other bones forming a structure not unlike the limb bones of tetrapods. Coelacanths are the most celebrated of lobe-finned fishes, although they are not the closest relatives of land vertebrates. These distinctive fish were long known as fossils, but only in rocks older than 66 million years, when they were thought to have become extinct. In 1938, however, a living coelacanth turned up among the haul of a fisherman working off the coast of South Africa, rendering conclusions about extinction premature. A second species was later discovered in waters near Sulawesi, in Indonesia. Coelacanths unmistakably have the lobed structure that marks them as vertebrate relatives, but equally unmistakably, they are fish, living their lives in water.

Closer tetrapod relatives are the lungfish, half a dozen species of freshwater fishes that not only have lobed fins but can breathe using primitive lungs related evolutionarily to swim bladders, bulbous organs in fish that are widely used to maintain buoyancy but also supply oxygen to the heart. Although still recognizably fish, lungfish show clear adaptations for life on land; only one species retains the ability to breathe using gills alone. The morphological gap between fish and tetrapods, however, remains large. Like plants, vertebrates required evolutionary transformation to colonize land. Not only did they need lungs to obtain oxygen from air, but they required structural reorganization of their skulls, ribs cage, and limbs to eat, breathe, and move about in terrestrial environments.

Fish capture food largely by sucking it into their mouths, and they obtain oxygen by gulping water and forcing it across their gills. Because of this, fish skulls are complex but flexible structures. On land, vertebrates obtain food by biting and oxygen by breathing air. As a result, the anatomy of the skull has been modified to form a stronger, more rigid structure that facilitates biting and air intake. These changes also adapted the palate for sound, with long-term behavioral consequences. Adaptation for breathing can also be seen in the evolution of the rib cage, long bones that extend from the backbone, supporting the muscles required to expand and contract the lungs. Moreover, in fish, bones that make up the shoulder girdle are continuous with the skull, largely serving to shape the sleek body that facilitates movement in water. Propulsion is supplied largely by muscles along the body and tail. Both structural support and locomotion of land vertebrates required muscular limbs attached to pronounced pelvic and shoulder girdles (now distinct from the skull, separated by a true neck) that anchor the muscles.

Remarkably, a series of fossils deposited between about 380 to 360 million years ago preserves a compelling record of this transition. Skeptics of evolution sometimes claim that fossils don't preserve evolutionary intermediates, but these folks haven't met *Tiktaalik* (Figure 38). Discovered in circa 375-million-year-old rocks from arctic Canada, *Tiktaalik* had the overall body organization of a lobe-finned fish, breathing via gills and covered in scales, but it had a flattened skull reminiscent of crocodiles. Its fins were built on the lobe-fin model, but with bone modifications that prompt thoughts of an elbow and wrist. The shoulder girdle was separated from the skull by a neck and appears to have supported muscles required for limb-like locomotion and body support. Features of the skull also suggest that, like living lungfish, *Tiktaalik* could breathe air by means of lungs.

Was *Tiktaalik* a fish, or was it a tetrapod? It isn't easy to tell, and that's the point. This remarkable fossil and others like it record the transition from water to land, with different features evolving on different timescales. Although still an aquatic animal, *Tiktaalik* was probably able to move and support itself in shallow water and, perhaps, adjoining land using its limb-like fins. Also, it could breathe air and capture prey in its jaws. Fossil trackways preserved on the surfaces of sedimentary beds provide independent evidence that by the late Devonian Period, vertebrates had begun to colonize dry land.

WHEN THE CAMBRIAN DIVERSIFICATION of marine life began, Earth's continents were engaged in global diaspora, as upwelling within the mantle fractured a late Proterozoic supercontinent and dispersed its pieces. But on the Earth's spherical surface, what goes around must come around, and so by Rhynie time, the continents had begun to amalgamate once again to form the single supercontinent called Pangaea. Landmasses collided over millions of years, pushing up mountains preserved today as gentle rises, their faulted and folded rocks

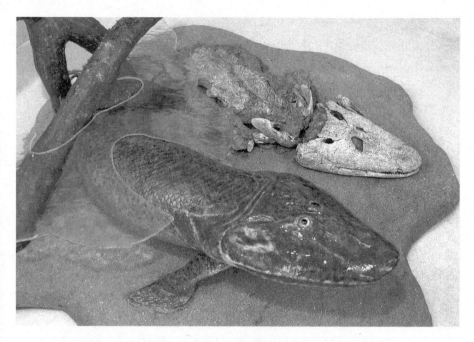

FIGURE 38. *Tiktaalik*, a 375-million-year-old fossil (reconstructed on left) that exhibits features intermediate between those of fish and land vertebrate animals. *Courtesy of Neil Shubin, University of Chicago*

visible in quarries and road cuts. Pangaea's assembly was complete by about 300 million years ago, but, driven by continuing mantle convection, the perpetual dance of the continents would fracture it into pieces again by 175 million years ago (see chapter 2).

How did life's conquest of land affect the planet? Soil is one product of this colonization. We commonly think of soils, if we think of them at all, as the physically altered surface of the planet. But soils, perhaps Earth's greatest resource for humans, reflect, once again, the *interactions* between physical and biological processes, owing as much to roots and fungi, buried plant debris and earthworms, as they do to chemical weathering. In fact, the principal physical part of soil formation—chemical weathering—is itself enhanced by roots that penetrate into the subsurface, releasing organic acids as they go. Thus, as terrestrial ecosystems developed, fertile soils developed along with them.

Cuticle, lignin, sporopollenin, and other biomolecules synthesized by plants resist bacterial decay, facilitating their burial and preservation in sediments. This new wrinkle on the carbon cycle should have had two distinct consequences. More burial of photosynthetically generated organic matter should have increased the transfer of carbon from CO_2 in the atmosphere to organic molecules in sediments, cooling Earth's climate. And, as organic carbon buried is organic carbon not respired using oxygen, enhanced organic burial should have increased O_2 levels in the atmosphere. Consistent with these predictions, four distinct lines of chemical evidence suggest that on the timescale of early land plant evolution, atmospheric oxygen finally reached modern levels, spreading O_2 through the depths of the ocean, as well. And beginning near the close of the Devonian Period and accelerating rapidly during the ensuing Carboniferous, continental ice sheets again expanded across the southern continents, their sedimentary signatures preserved in southern Africa, South America, India, Australia, and Antarctica, all part of a single late Paleozoic landmass.

As ice crowned the poles, swamps spread across equatorial lowlands, at the time including North America, Europe, and parts of China. Much of the coal that sustained the Industrial Revolution (and fuels global warming) formed from plant debris buried in these ancient wetlands. Biologically, this was a time of giants—not yet dinosaurs, but dragonflies with wingspans up to 28 inches (70 centimeters) and millipedes seven feet (2 meters) long. Horsetails, a group of plants today represented by only fifteen species of mostly small stature, included trees more than 30 feet (10 meters) high. And club mosses, again mostly limited to small ground cover in modern landscapes, reached heights of 100 feet (30 meters) in tropical Carboniferous wetlands. The coal of West Virginia, Kentucky, and Illinois is largely the compressed remains of these extinct giants. Ferns and seed plants diversified as well, mostly extinct taxa but including, among other things, the ancestors of modern conifers. The wetlands, however, were not to persist. The mountains that rose during late Paleozoic continental collisions altered patterns of atmospheric and ocean circulation, draining the wetlands and dooming the distinctive species they supported. Among both plants and tetrapods, new ecosystems were taking shape, ones that would come to include the most iconic of all extinct organisms, the dinosaurs.

YEARS AGO, when I actually qualified for the distinction, I participated in a conference for young scientists. Among the colleagues I met was Maria Zuber, then a budding planetary scientist and now a renowned expert on the moon and distant planets. At the close of the first day, Maria called home to catch up with her young son, telling him that she had spent much of the day talking to paleontologists. Excited, he asked whom she had met. "Oh, a couple of people," Maria replied. "Andy Knoll,

Simon Conway Morris." Never heard of them. Told that these no-names worked on early life, Maria's boy responded in evident sympathy. "Don't worry, Mom," he consoled. "Maybe next time you'll meet some dinosaur people."

Dinosaurs. *Brachiosaurus, Triceratops, Tyrannosaurus rex.* You know their names, or at least you did when you were eight. Viewed within the full panoply of Earth and life, their hegemony was brief—less than 4 percent of our planet's history. And their impact on the Earth itself pales before that of cyanobacteria. But during the Jurassic and Cretaceous periods the ecological dominance of dinosaurs was absolute, and the shapes they evolved are without parallel in the history of life.

So, what were dinosaurs? What features underpin their ecological success? And why did some of them attain such impressive size? Let's begin by placing dinosaurs within the world they inhabited.

The earliest land vertebrates were predators, and perhaps scavengers, but within 50 million years, tetrapod diversity came to include both carnivores and herbivores, divided between amphibians and amniotes, the group that today includes reptiles, birds, turtles, and mammals. As described in the next chapter, the Paleozoic Era ended in catastrophe, but as terrestrial ecosystems revived in the Mesozoic Era (252–66 million years ago), both vertebrates and vegetation took on a more modern aspect. Now the dominant trees and shrubs were conifers, ginkgos and other seed plants, with diverse ferns in the understory. Flowering plants, which dominate most land ecosystems today, radiated only late in the era, their earliest fossils a bit more than 140 million years old.

Early Mesozoic tetrapod diversification also gave rise to groups we still see today. The earliest known true mammals, turtles, lizards, and frogs all occur in rocks of the Triassic Period (252–201 million years ago), along with pterosaurs (the first winged vertebrates), dinosauromorphs (the earliest true dinosaurs and their close relatives), and other now-extinct groups. In many Triassic landscapes, the most diverse and abundant vertebrates were large agile reptiles, some bipedal and others running on all four legs, some with long toothy snouts and some pug-nosed, some carnivorous and others plant eaters. Dinosaurs? Actually, no. While dinosaurs existed in Triassic communities, they weren't particularly abundant or diverse. The aristocrats of Triassic landscapes belonged to a lineage today represented by crocodiles. Did dinosaurs eventually take over by dint of some superior adaptation? It doesn't look that way. The Triassic world emerged in the wake of catastrophe and ended that way, too. Dinosaurs rose to ecological prominence at least in part because they survived late Triassic environmental distress, perhaps as much by good luck as by good genes.

Even as a more modern biological world was taking shape on the continents, the physical Earth continued to change. Among the early records of Pangaea's breakup are the Palisades, volcanic rocks erupted during early fracturing of the supercontinent and now exposed in low cliffs along the Hudson River near New York City. The Atlantic Ocean opened in zipper-like fashion, from the equator toward the poles. And as both North and South America moved westward, oceanic crust of the Pacific plate subducted beneath them, giving rise to the Rockies and Andes. The southern continents also broke apart; Africa and India headed north, colliding with the underside of Eurasia to form the great mountains that extend from the Alps to the Himalaya. That is, the global geography know today was taking shape. It was a warm, largely ice-free interval, but, again, as plate tectonics redistributed landmasses and pushed up mountains, the seeds were planted for a new ice age, still far in the future.

NOW WE CAN RETURN to basic questions about dinosaurs. The definition of a dinosaur is actually fairly prosaic. Beginning in the early 1800s, paleontologists discovered a number of enormous fossils, distinct from any tetrapod alive today, and gave them an evocative name: dinosaurs (from the Greek for "terrible lizards"). Today the group is defined in terms of genealogy: dinosaurs include the last common ancestor of these first-found behemoths and its descendants. Fortunately, this definition aligns pretty well with the image conjured up when someone mentions the term, but as we shall see, it has a surprising consequence.

When we think of dinosaurs, most of us envision gigantic creatures, with even the herbivores looking more than a little forbidding. That's broadly correct, although the smallest known dinosaurs weighed only about 15 pounds (7 kilograms), the size of a miniature schnauzer. A recent compilation of body size among vertebrate species shows that for most groups, be they mammals, birds, amphibians, or fish, size distribution is weighted toward smaller bodies, with a long, low distributional tail extending to larger species—lots of rodents but not many elephants. Dinosaurs, however, are different; their sizes are actually skewed toward large bodies.

So as any eight-year-old could tell you, most dinosaurs really were big. But why should this be the case? Why were dinosaurs different from other tetrapods that have roamed the

Earth through time? There is no consensus on the answer, but German paleontologist Martin Sander and his colleagues have articulated a hypothesis that strikes me as sound.

The first dinosaur giants, and indeed, the largest dinosaurs of all time, were the sauropods, long-necked herbivores whose largest representatives, the titanosaurs, reached lengths up to 120 feet (37 meters) and weights of 70 to 90 tons. (The magnificent specimen on display at New York's American Museum of Natural History is cleverly mounted so that its head sticks out into the hallway, underscoring its immensity [Figure 39].) Sander and his colleagues draw particular attention to those long necks.

The remarkable necks of sauropods enabled them to reach food not available to other plant eaters and to forage over large areas with minimal motion—the larger they grew, the more effective they became at garnering food resources. Long necks were possible because sauropods have very small heads—the necks of sauropods couldn't have supported heads the size of those on hadrosaurs or tyrannosaurs. Small heads, in turn, were possible because, unlike obedient children, sauropods didn't chew their food. They simply—and rapidly—nipped and stripped branches, swallowing leaves and seeds more or less whole.

THE TITANOSAUR

Unlike crocodiles, dinosaurs had a bird-like respiratory system, facilitating the efficient transport of oxygen through a mammoth body and, importantly, leading to vertebrae with numerous air cavities that lightened the weight of the neck. Also, sauropods had high metabolic rates, enabling them to grow rapidly, a necessity for species whose adults could be 100,000 times larger than hatchlings. Today animals are usually classified as warm-blooded, maintaining high body temperature by burning a lot of calories, or cold-blooded, relying on the environment to regulate body temperature. Warm-blooded mammals and birds use a lot of the energy they take in as food to maintain those high internal temperatures. It appears that while dinosaurs weren't warm-blooded in the sense we associate with birds and mammals today, they were able to maintain elevated body temperature in a distinct way that promoted efficient metabolism, while devoting more of their food intake to growth. The key, not surprisingly, was size. As an animal grows larger, the heat it generates increases as a function of its volume (the cube of length), while body heat is dissipated as a function of surface area (the square of linear dimensions). Thus, at the large sizes

FIGURE 39. *Patagotitan mayorum*, a gigantic titanosaur skeleton on display in the American Museum of Natural History, New York. From snout to tail, the skeleton is 122 feet (37 meters) long. © *American Museum of Natural History/D. Finnin*

attained by dinosaurs, high internal temperatures could be maintained passively. Recent support for this view comes from chemical analyses of sauropod bones, which indicate a body temperature of 97–100 degrees Fahrenheit (36–38 °C), much like living mammals.

For sauropods, size provided a strong defense against predators (elephants seldom fear leopards). In response predators grew larger, setting off an evolutionary arms race across dinosaurs as a whole. In consequence, the ecological and physiological sweet spot on land became that of the terrible lizards. Early mammals lived in the same communities as dinosaurs, but were unable to attain similarly large size. To survive, most simply stayed out of the dinosaurs' way, living nocturnally, or in trees or burrows, as many mammal species do today. For those who favor David over Goliath, it is worth noting that at least some of these early mammals ate dinosaur eggs.

WE LIKE TO THINK of dinosaurs as extinct, but if we accept the definition of dinosaurs offered earlier, that's not true. You have living dinosaurs in your own backyard—finches, warblers, and sparrows. The idea that birds are descended from dinosaur ancestors goes back a century and a half to T. H. Huxley, Darwin's most tenacious supporter. In 1868, Huxley wrote, "The road from Reptiles to Birds is by way of Dinosauria . . . wings grew out of rudimentary fore limbs." Huxley drew particular attention to anatomical similarities between the skeletons of birds and that of *Coelophysis*, a small dinosaur found in late Triassic and early Jurassic rocks.

Fossils of intermediate character once again bolstered the case. In 1855 and again in 1861, two remarkable fossils were excavated from a limestone quarry in Bavaria. Christened *Archaeopteryx*, the fossils reveal an overall skeletal structure much like that of contemporaneous small dinosaurs, but with forelimbs extended to wing-like dimensions (Figure 40). The skull was modified to resemble the bills of birds, although the jaws were still lined with teeth. More astonishing, *Archaeopteryx* was covered in feathers. (The iconic Berlin specimen of *Archaeopteryx* can be seen at the Humboldt Museum, displayed prominently and protected by bulletproof glass like *Mona Lisa* in the Louvre.) As *Tiktaalik* did for the fish-tetrapod transition, *Archaeopteryx* reveals both where it came from, in terms of evolution, and where it was heading. In recent decades, the dinosaur-bird connection has been strengthened by dozens of new discoveries from Cretaceous beds in China, showing, among other things, that the dinosaurs most closely related to birds already sported

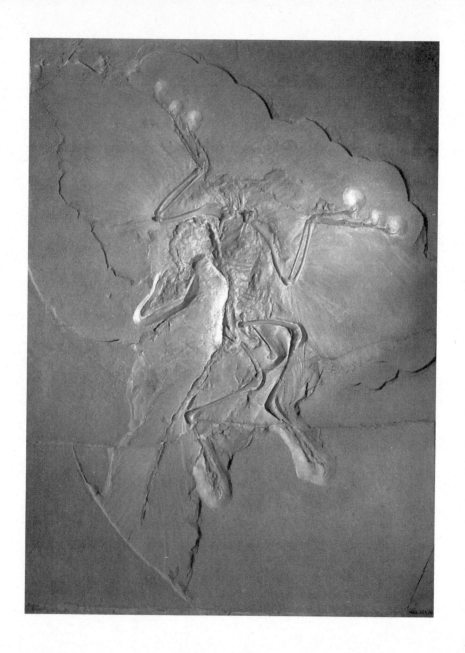

feathers. Preservation of pigment molecules even allows us to reconstruct color patterns in these avian forerunners, providing a new scientific answer to the old riddle "What's black and white and red all over?" While early proto-birds may have used elongated forelimbs to catch prey, they eventually evolved the ability to glide and then fly actively—skeletal and muscular modifications needed to fly can be seen in the diverse fossils from China. Flight provided a new realm to conquer—the air. Pterosaurs got there first, and recently it has been shown that other small dinosaurs evolved wings independently of the birds, but birds did it best, commanding the sky (to be joined by bats much later) and, importantly, surviving renewed environmental catastrophe 66 million years ago. So when you talk to your parrot, admire the grace of eagles, roast your chicken, or shoo crows from your garden, give birds the respect they are due—they are the survivors of the mighty dinosaur clade.

FIGURE 40. *Archaeopteryx lithographica*, a remarkable fossil that links dinosaurs and birds. This is the original specimen displayed at the Museum für Naturkunde in Berlin. © *H. Raab (User: Vesta)/source: https:// commons.wikimedia.org/wiki/File:Archaeopteryx_lithographica_(Berlin_ specimen).jpg*

7

Catastrophic **Earth**

EXTINCTIONS RESHAPE LIFE

NEAR THE MEDIEVAL TOWN of Gubbio, Italy, a narrow gorge slices deep into the Apennine Mountains. Rocks along the canyon walls may strike casual observers as monotonous, layer after layer of fine-grained limestone deposited long ago on a deep seafloor. The limestones are full of fossils—in fact, they consist mainly of the tiny calcium carbonate skeletons of protozoans called foraminiferans and microscopic algae called coccolithophorids—but because of their minute size, these remains aren't apparent in the rock face. There is one curious feature, however, if you look carefully in just the right place. Above hundreds of meters of limestone and beneath many more similar beds, there is a half-inch (1 centimeter) layer of clay with no carbonate minerals at all (Figure 41). Take the limestones back to the lab to study, bed by bed, under the microscope and you'll find another puzzle: few of the microfossil species found below the clay layer persist above it.

The clay layer at Gubbio marks the boundary between the Cretaceous and Paleogene periods—and the Mesozoic and Cenozoic eras—a fence in time at 66 million years ago that separates distinctly different biotas both on land and in the sea.

In marine rocks, the microfossil species used to reconstruct Mesozoic time disappear, seemingly in an instant. Ammonites, squid relatives that were among the most abundant and diverse carnivores in Mesozoic oceans, also disappear en masse, as do myriad other species. On land, the long-dominant dinosaurs breathed their last. All, it turns out, at the exact moment in time represented by the Gubbio clay layer.

In the late 1970s, geologist Walter Alvarez traveled to Gubbio to study the magnetic properties of its thick limestone beds, and that singular clay layer caught his attention. How much time might it represent? Walt raised the question with his father, physicist and Nobel laureate Luis Alvarez. The problem was easy, replied Alvarez *père*; tiny micrometeorites constantly rain down through the atmosphere and do so at known rates. These heavenly messengers contain elements like iridium that are rare in Earth surface materials, so if Walt measured the

FIGURE 41. The Cretaceous-Paleogene boundary in Gubbio, Italy, where Walter Alvarez developed the case for mass extinction by meteorite impact. White limestones to the lower right were deposited toward the end of the Cretaceous Period; they contain diverse skeletons of tiny foraminiferans and coccolithophorid algae. The reddish limestones in the upper left formed at the beginning of the Paleogene Period; they contain only a few foram and coccolithophorid species. Separating them is a thin layer of fine mudstone, at the top of the white zone, much sampled by curious geologists. *Andrew H. Knoll*

amount of iridium in the clay layer, he could calculate the time it took to accumulate. With chemists Frank Asaro and Helen Michel, he did just that, calculating from its high iridium content that the clay layer must have taken millions of years to form, an answer that Walt knew had to be wrong—marvelously, spectacularly, and instructively wrong, as it turned out.

If the iridium enrichment in the clay didn't reflect slow rates of accumulation over long timescales, the alternative was a lot of iridium deposited quickly, most likely by the impact of a large meteorite; Alvarez and his colleagues calculated the required size to be nearly seven miles (eleven kilometers) in diameter. The planetary effect of such a collision would be catastrophic, providing a mechanism for the extinction of dinosaurs and all the other plant, animal, and microscopic species that never saw dawn break on the Paleogene Period.

Published in 1980, the Alvarez team's paper was a sensation, generating supporters and skeptics in equal measure. The controversy lasted for a decade or so until accumulating data tipped the scales decisively in the Alvarez' favor. (As an aside, when Walt visited Harvard in the late 1980s, he stayed at our home. I introduced him to my daughter Kirsten, then aged four, telling her that "Mr. Alvarez is interested in dinosaurs." Kirsten was immediately entranced, so I ventured a step further. "Are there any dinosaurs living today?" I asked. "No, silly," Kirsten responded, her tone of voice clearly lamenting my ignorance, "A meteorite killed them." Walt jumped up from the couch, his arms shooting into the air like a referee signaling a touchdown. If kids accepted the story, scientists were sure to follow.)

Like other questions in science, the Alvarez hypothesis wasn't settled by a vote. The hypothesis made predictions about other features that should be preserved in the rock record, and geologists all over the world set out to look for them. The iridium anomaly showed up in boundary rocks around the world, but not in older or younger beds. And soon, distinctive minerals called shocked quartz were discovered in rocks deposited at the same point in time. Shocked quartz forms only under conditions of transiently high temperature and pressure, conditions generated by the impact of a large meteorite. And in time, the true smoking gun was identified: a meteorite crater some 125 miles (200 kilometers) in diameter, formed at just the right time and now buried beneath younger sediments in the Yucatan Peninsula. Nearly 170 million years of dinosaur evolution ended in cataclysm.

IF YOU ASK a paleontologist what fossils contribute to our understanding of evolution, she'll probably point to long-vanished organisms like dinosaurs, trilobites, and giant club mosses that expand our appreciation of biological possibility, before homing in on mass extinctions and their profound consequences for life. It wasn't always this way. In 1944, George Gaylord Simpson, paleontology's leading contributor to the so-called Neodarwinian evolutionary synthesis of the mid-twentieth century, wrote a highly influential book called *Tempo and Mode in Evolution*. In this volume, Simpson argued that evolutionary patterns defined by fossils reflect population genetics acting over long intervals of time. His arguments were straightforward and compelling—after all, the point of the Neodarwinian synthesis was to establish population genetics as the mechanism underlying natural selection and hence evolutionary change through time. By focusing so resolutely on population genetics, however, Simpson missed a key geologic lesson for evolution. Earth isn't a passive platform on which dynamic populations evolve; our planet is just as dynamic as the populations it supports, with environments continually changing on scales that range from local and transient to long-term global transformation. And when environmental disruption subjects the biota to a short, sharp shock, species and even ecological structure can collapse. Population genetics certainly underpins the *origin* of species, but the *persistence* of species is commonly adjudicated by Earth's environmental dynamism. As previous chapters have hinted and end-Cretaceous events make clear, the biological diversity we see around us today reflects mass extinction and environmental change every bit as much as it does population genetics. Mammals radiated on the Cenozoic Earth not just because of population genetics but also because some of them survived end-Cretaceous catastrophe while dinosaurs didn't.

The Alvarez hypothesis did much to focus paleontological thinking on mass extinction, while further impetus came from another project that took shape about the same time. When I was a graduate student in the 1970s, my friend and fellow student Jack Sepkoski started to tabulate fossil diversity through time. Jack wasn't the first to try this, but his perseverance and attention to detail enabled him to put together a remarkable database of the first and last appearances of every order, family and, eventually, genus of marine animals found in the fossil record. (Jack stayed away from tabulating species, correctly intuiting that the record at that level of detail would be prone to biases related to sediment abundance and the habits of collectors.) Jack's data showed that the course of biological diversification never did run smooth. Animal diversity expanded in the Cambrian and Ordovician periods, but then plummeted near the end of the Ordovician. It rebounded, only to fall once more during the later Devonian Period, and then repeated the cycle three more times, including the end of the Cretaceous Period. In all, Earth's biota has endured five mass extinctions during the past 500 million years, along with half a dozen lesser extinction episodes (Figure 42).

At first, it seemed like the Alvarez hypothesis might provide a general explanation for Sepkoski's portrait of fluctuating

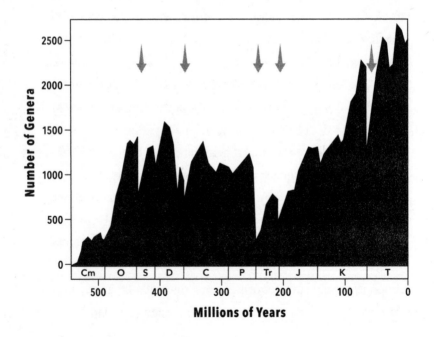

diversity. Perhaps big meteorites caused big extinctions, and smaller impacts drove smaller extinctions. Simple, yes, but as it turns out, wrong. Only the end-Cretaceous extinction is reliably associated with meteorite impact.

The largest known mass extinction occurred not at the end of the Cretaceous Period, but rather 252 million years ago, at the close of the Permian Period, when more than 90 percent of marine animal species disappeared. (It might seem a coincidence that these two great extinction events fall at the boundaries between the eras of the Phanerozoic Eon, but, of course, it isn't a coincidence at all. Nineteenth-century paleontologists established the geologic timescale on the basis of fossils, and the pronounced paleontological changes at the ends of the Permian and Cretaceous periods marked these as natural points for subdividing Earth history.)

End-Permian biological catastrophe is clearly inscribed in rocks exposed on a mountainside at Meishan, China (Figure 43). The site is easy to find because the regional government built

FIGURE 42. A compilation of the genus-level diversity of marine animals through time, painstakingly developed by Jack Sepkoski. Arrows point to five moments during the past 500 million years when diversity plummeted rapidly—the "Big Five" mass extinctions. *Source: Sepkoski's Online Genus Database*

a garish geo-park to preserve, exhibit, and exploit the local geology. But human embellishments aside, the rocks at Meishan tell a chilling story. Limestones near the base of the mountain bristle with the fossils of late Permian marine animals: brachiopods, bryozoans, echinoderms, large skeletonized protists, and more. Had you been able to swim along the late Permian coast, you would have observed diverse animals, seaweeds, and protozoans distributed across the shallow seafloor. Halfway up the section, however, these fossils simply disappear. All of them. At a point the thickness of a knife blade. The fossils never reappear in younger rocks. Instead, as we continue up the mountainside, we see only a few small fossils, mostly clams and snails.

When I first saw all this at Meishan, I felt a surprising sense of existential loss—an exuberance of life struck down rapidly and forever. But what happened? The first step toward an answer comes from thin layers of volcanic ash interspersed among Meishan limestones; ash beds just above and immediately be-

FIGURE 43. The Permian-Triassic boundary exposed in Meishan, China. The massively bedded rocks in the lower right are later Permian limestones rich in fossils. Above them, the rocks turn abruptly to fine-grained limestones that contain few fossils. Some 90 percent of marine animal species went extinct at the point in time marked by the change in sedimentary rock type. *Andrew H. Knoll*

low the extinction horizon yield ages of 251.941±0.037 and 251.880±0.031 million years, respectively. These precise ages are key because they coincide with the timing of a startling geologic event half a continent away, the Siberian Traps.

In geologic parlance, a "trap" connotes an accumulation of basalt or other dark-colored volcanic rocks, commonly exposed bed after bed in a way that suggests stairs (*trappa* in Swedish). Located east of the Ural Mountains, the Siberian Traps record an immense outpouring of basaltic rocks, not unlike those that flow across the island of Hawaii. While the Siberian Traps may be similar in type to volcanic eruptions observable today, their dimensions are staggering. The surviving aerial extent of the traps is about three million square miles (seven million square kilometers), roughly the size of Australia. Commonly more than 8,000 feet (2,500 meters) thick, the traps have an estimated volume of a million cubic miles (four million cubic kilometers), a million times greater than any volcanism ever witnessed by humans or our close relatives. Careful radiometric dating shows that most of this immense pile erupted at just the same time as the extinctions recorded at Meishan.

How do we connect volcanism in western Asia with biological catastrophe well illustrated in China but observed globally? While widespread, the Siberian Traps didn't cover the Earth or anything like it, so global extinctions don't simply reflect inundation by lava. We have to ask how massive volcanism would have impacted global environments. As volcanoes spread lava locally, they inject large amounts of gas into the atmosphere, especially carbon dioxide, a geologic actor whose effects on climate we've seen before. End-Permian volcanism rapidly increased the CO_2 content of the atmosphere and oceans by severalfold.

More than twenty years ago, my friend Richard Bambach and I became intrigued by end-Permian mass extinction. Other paleontologists had examined the extinction record before, seeking meaningful patterns in the geographic, environmental, or taxonomic features of victims and survivors. In contrast, Dick and I looked to physiology, the biological interface between organisms and their surroundings. In particular, we asked ourselves how life would be affected by the massive injection of carbon dioxide into the atmosphere. This was before we knew much about Siberian volcanism, and to be honest, our motivating model for the extinction was wrong. Nonetheless, the results proved illuminating. Months in the library taught us what physiologists had learned from decades of laboratory experiments. In high concentrations, carbon dioxide is bad news for many organisms, affecting their environment and physiology in equal measure. But not all species respond to the same degree—some are relatively tolerant, while others can be especially vulnerable. We developed a list of anatomical and physiological traits that can reasonably be inferred from fossils and used these to divide the late Permian marine fauna into two groups, one predicted to be more tolerant to rapid CO_2 increase and the other more vulnerable. The actual record of end-Permian extinction and survival fit the predictions remarkably well, identifying carbon dioxide and other volcanic gases as the link between physical calamity and biological catastrophe.

Siberian Trap volcanism injected vast quantities of carbon dioxide into the atmosphere, which, because of the greenhouse properties of CO_2, led to global warming. (Because the Siberian Traps flowed over vast areas of accumulating peat, methane [CH_4] may also have formed from heated organic matter, exacerbating the greenhouse effect.) Because warming decreases the amount of O_2 that can be mixed into seawater, the seas became oxygen-poor, especially in subsurface waters not in direct contact with the atmosphere. And the mixing of emitted CO_2 into seawater also decreased its pH—what we now call "ocean acidification." German physiologist Hans Otto Pörtner, a leader in efforts to understand the biological consequences of twenty-first-century global change, calls global warming, ocean acidification, and oxygen depletion "the deadly trio." Not only is each factor capable of harming the biota individually, they all occur together in the Earth system and their effects are synergistic—each factor makes the effects of the others worse. Hypercapnia, the direct physiological effects of increased CO_2, was also in play. For example, at high carbon dioxide levels, proteins that transport O_2 through the body can bind instead with CO_2, hindering oxygen metabolism.

CO_2's concatenated environmental and physiological effects should be most pronounced in animals that make massive carbonate skeletons, but have only a limited physiological capacity to modify the fluids from which those skeletons are precipitated—corals, for example. In contrast, animals with high metabolic rates—exposing them to high internal CO_2 on a daily basis—should be more tolerant, as should be animals with gills or lungs for gas exchange and a well-developed circulatory system. With this in mind, we might predict that mollusks, fish, and arthropods would prove relatively tolerant. As volcanism raged at the end of the Permian Period, biology was indeed destiny in the oceans. All Paleozoic corals disappeared—the corals in today's oceans reflect the later Triassic evolution of skeletons in sea anemones that survived the extinction. Brachiopods, physiological couch potatoes that were among the most wide-spread and diverse animals on the Permian seafloor, lost nearly all of their diversity. On the other hand, clams and snails survived relatively well. Consistent with physiological predictions, fish experienced relatively few extinctions, and decapod crustaceans, epitomized today by the shrimps, crabs, and lobsters on our dinner plates, actually diversified from the Permian into the Triassic Period.

On land, most animals and plants felt the effects of global change, but perhaps because populations on land were not subject to ocean acidification or oxygen loss, and perhaps because terrestrial species are more tolerant of temperature change, the long-term effects appear to have been more modest than those that so altered the sea. All in all, the patterns of ecology and diversity that had characterized Earth's oceans for more than 200 million years collapsed—not because of extraterrestrial influence but because hot magma from the mantle erupted violently across the Siberian landscape. Diversity rebounded during the Triassic Period, but now with different groups contributing to a distinct ecology. Mass extinction ended the Paleozoic Era and ushered in the Mesozoic, just as end-Cretaceous catastrophe closed the books on the Mesozoic and potentiated our Cenozoic world.

DESPITE THEIR APOCALYPTIC DIMENSIONS, the Siberian Traps are not a geologic singularity. Spurred by focused heat from the mantle, large volumes of lava have erupted across the landscape or seafloor eleven times in the past 300 million years, providing a mechanism to explain at least one other mass extinction and several smaller events. In the wake of end-Permian extinction, marine life diversified again during the Triassic Period (252–201 million years ago), giving rise to new and distinct ecosystems over an interval of several million years. But the Triassic Period ended as it began, as massive amounts of lava erupted along an arc that extends from Fingal's Cave off the west coast of Scotland and the Palisades of New York to black cliffs in Morocco's Atlas Mountains and extensive flows now buried by Amazon rain forest. And again, biological diversity plummeted. The selectivity of end-Triassic extinction and survival in the oceans mirrors that of the Permian, with reefs hit particularly hard. An estimated 40 percent of all genera and up to 70 percent of species disappeared from the oceans, well below the numbers for end-Permian extinction, but still dramatic. On land, the Triassic diversification of vertebrates was curtailed by volcanism and slightly earlier climate perturbation; as noted in the preceding chapter, diverse crocodilians, dominant components of later Triassic landscapes, became extinct, while dinosaurs and mammalian ancestors survived, giving rise to the great and the small of later Mesozoic ecosystems.

The later Mesozoic geologic record documents several additional moments in time when large portions of the deep ocean became oxygen-starved for thousands of years. At least two of these events correlate with both massive volcanism and elevated extinction, one about 183 million years ago and a second 94 million years ago. An estimated 15–20 percent of all marine genera disappeared during these "minor" extinctions. There is even massive volcanism at the end of the Cretaceous Period. Some scientists believe that India's Deccan Traps volcanism influenced the course of end-Cretaceous mass extinction, either setting the stage for the impact by perturbing the environment or exacerbating the effects of impact by spewing great volumes of gases into the air. Some radiometric dates for the lavas favor the former view, while others support the latter. Fortunately, massive volcanism can impart a chemical fingerprint to sediments, and this shows that Deccan volcanism began before the extinction. Regardless of geologic context, however, fossils get the final word on causation. Observed patterns of end-Cretaceous extinction and survival bear little resemblance to those of extinctions reliably associated with massive volcanism, stressing the importance of meteorite impact in closing the door on the Mesozoic world.

OUR TWO FINAL MASS EXTINCTIONS, both from the Paleozoic Era, are distinct in terms of both cause and effect. In chapter 5 we learned that Cambrian and Ordovician diversification fashioned diverse ecosystems in the oceans. That diversity, however, collapsed 445 million years ago, near the end of the Ordovician Period. The extinction coincides with a relatively brief (two million years) but pronounced ice age, centered on the Southern Hemisphere. Something approaching half of all marine animal genera became extinct, but with only limited tears in the ecological fabric; seafloor communities as the world recovered looked much like those before the extinction event. Life in the water column was affected more, with sharp drops in diversity among trilobites and early vertebrates.

It's a bit puzzling that end-Ordovician mass extinction should coincide with widespread glaciation, as the Earth has been locked in an ice age for the last 2.6 million years (chapter 8), and in the marine realm at least, only modest extinction has been documented. Why, then, should the end-Ordovician world have been different? One distinction concerns sea level. The water in glaciers comes mostly from the oceans, so as ice expands, sea level drops—about 425 feet (130 meters) in the case of our most recent ice age and not too different at the end of the Ordovician Period. If sea level is low at the outset, as it was when glaciers expanded most recently, the resulting loss of habitable seafloor is relatively modest. If, however, large glaciers expand when sea level is high and much of Earth's low-lying land is flooded by shallow seas, the resulting drop in sea level will drain the shallow seas that flooded continents, eliminating a large swath of the world's shallow seafloor, along with its inhabitants. That is what happened during the Ordovician Period.

Another issue concerns geography. As climate changes, populations can move to more hospitable environments if migration routes are available. As ice expanded 2.6 million years ago, plant species in eastern North America were able to relocate closer to the Gulf of Mexico, ensuring their survival. In contrast, plants from northern Europe, blocked by the Alps, suffered significant extinction. In shallow oceans, extinctions were greatest where migration routes were limited—in Florida, for example, where deep seas limited migration to warmer climes. (Ask yourself where polar bears can migrate as the current Earth warms.) Rising equatorial mountains and deep seas may likewise have inhibited end-Ordovician migration. Neither limits to migration nor large-scale habitat loss should bias patterns of species loss in ways similar to the end-Permian catastrophe, perhaps helping us to understand why despite major loss of species, ecological patterns largely persisted through end-Ordovician mass extinction.

The least well understood event of Sepkoski's "Big Five" occurred during the later Devonian Period, when diversity dropped over a protracted interval (393–359 million years ago). Brachiopods and other organisms nestled on the seafloor went first, then reef builders, and finally early cephalopods jetting through the water column. Curiously, Devonian diversity decline appears to reflect low rates of origination as much as it does extinction, leading Dick Bambach and me to christen this event a "mass depletion" rather than a canonical mass extinction. That diversity loss relates to rates of origination has been borne out in several studies, but why this should be so remains a topic of ongoing research.

CAN WE DRAW GENERALIZATIONS about Earth's extraordinary and repeated record of mass extinction? Common causation is out the window—different events relate to meteorites, ice ages, and massive volcanism. Neither is ecological impact generalizable—extinction events show contrasting ecological impacts, with ecosystem disruption not closely mirroring severity of species loss. Where we do find common ground is that in each case, environmental disruption was rapid; the *rate* of environmental change was as important as its *magnitude*.

When environmental change is slow, populations can adapt to their shifting circumstances, but when it is fast, adaptation may be challenging, leaving migration or extinction as the only available options. Mass extinctions reflect transient but profound environmental disruption, driven by mechanisms within the Earth or elsewhere in our cosmic neighborhood. While the timescale of mass extinction is short, however, rebuilding diversity is more protracted—fossils tell us that recovery from major extinctions takes a long time—hundreds of thousands or even millions of years.

Mass extinctions have clearly played a major role in shaping evolutionary history. The modern world is full of mammals in part because dinosaurs became extinct. Fish in the open ocean radiated only after end-Cretaceous mass extinction eliminated the ammonites. Reefs today contain modern corals, mollusks, and crabs not so much because they outcompeted the tabulate corals, brachiopods, and trilobites of ancient reef systems, but because mass extinction decimated those groups. When you trek through a rain forest or snorkel above a coral reef, you might reflect that you're surveying the survivors of Earth's repeated mass extinctions.

Could it happen again? Large meteorites and massive volcanoes are rare, but there is no reason to think that we've seen the last of them. In 43 B.C., the eruption of an Alaskan volcano caused harsh winters and widespread crop failure in Europe, contributing to the demise of the Roman Republic, and as recently as 1815, Mount Tambora in Indonesia blew its top, killing thousands of people locally and precipitating a "year without a summer" as far away as New England. And then there's Pompeii. (Naples itself sits atop debris from an eruption nearly 4,000 years ago.) Large meteorite impacts are decidedly rarer, but the Tunguska event, a massive explosion in 1908 that leveled some 80 million trees in a (fortunately) sparsely populated part of Siberia, is attributed to the aerial disintegration of an incoming comet or meteorite.

Thankfully, volcanoes and impacts large enough to cause global devastation are rare on time scales of millions of years, so I wouldn't worry too much about them. Far more concerning is what you see as you walk down the street—a human population capable of altering Earth and life profoundly within the lifetimes of you and your children.

8

Human **Earth**

ONE SPECIES TRANSFORMS THE PLANET

AS THE LAST EMBERS of end-Cretaceous cataclysm cooled around 66 million years ago, our planet began a new chapter. Surviving plants and animals began to diversify almost immediately, establishing renewed and resilient ecosystems on land within a few hundred thousand years. Already balmy, Earth grew even warmer over the next fifteen million years, the greenhouse consequences of relatively abundant carbon dioxide in the atmosphere. Palm trees came to flourish in Alaska, while alligators slithered across the Canadian Arctic. With dinosaurs gone, mammals diversified in new ways, becoming dominant components of terrestrial communities. Of particular interest were tiny, tarsier-like creatures that lived in tropical trees and probably dined on insects—the first primates, our ancestors.

Through Cenozoic time, life and environments changed in concert. Continents continued the global diaspora begun earlier as the supercontinent Pangaea broke part. The Atlantic Ocean widened dramatically, and the Rockies, Alps, and Himalaya all rose majestically into the sky. Rising mountains increased weathering rates, pulling carbon dioxide from the atmosphere, while shifting plates reoriented seawater circulation in the oceans. In consequence, Earth eventually began to cool. Palms, alligators, and other warmth-loving species retreated from high latitudes, and grasslands began to replace forests in continental interiors. By 35 million years ago, glaciers began their spread across Antarctica.

Against this dynamic physical backdrop, primates radiated across landscapes, giving rise to a diverse assortment of lemurs, tarsiers, monkeys, and our own branch of the primate tree, the great apes. We pick up the story 6–7 million years ago, as global cooling began to quicken Earth's approach to another ice age. In Africa, open woodlands and grassy plains increasingly replaced forests as the continent's interior dried, and spurred by this shift in habitat, a new lineage of great apes diverged from its closest relatives, today represented by chimpanzees and bonobos. Called hominins, these new apes were broadly chimplike: small in stature, with a small brain, protruding snout, and long arms with elongate curving fingers to facilitate movement through tree tops. But the hominins differed from other great apes in one critical way. They could walk upright.

Among great apes, only humans walk erect, our posture and locomotion made possible by a series of anatomical adaptations, including a curved lower spine to balance the upright trunk, a reconfigured pelvis to facilitate muscles needed for walking, a vertically oriented neck that places the head square atop the body, and an arched foot with a prominent heel. You display these features and, to a degree, so did the earliest hominins. These ancestors are known from fragmentary skeletons in rocks as old as 6–7 million years, but our best insights come from a single well-preserved skeleton of a young woman discovered in 4.4-million-year-old rocks from Ethiopia. *Ardipithecus ramidus*, or Ardi for short, shows many features likely present in the common ancestor of humans and chimps: she was an adept climber, at home in the trees. But Ardi also foraged in open woodlands for fruit and other food. As Charles Darwin first suggested more than a century ago, bipedal locomotion freed up her hands for other functions, including, in time, the manufacture and use of tools. Bipedalism, then, set Ardi and her relatives on the road to us (Figure 43).

A bit after Ardi's time, a new group of hominins emerged. Called australopithecines, these apes resembled the earliest hominins, but key differences place them further along the evolutionary path to humans. We don't know their true diversity, but a dozen or so species have been recognized to date, all from Africa. Australopithecine bones are relatively common, but a single skeleton once again sheds unusual light on the group. Lucy is probably the most famous of all pre-human hominins.

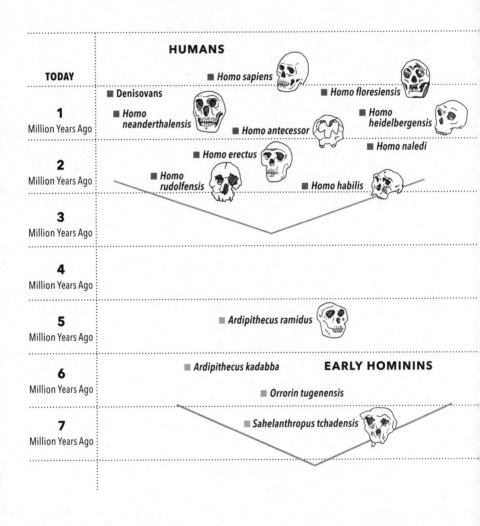

HUMANS

TODAY

Homo sapiens

■ Denisovans

1
Million Years Ago

■ **Homo neanderthalensis**

■ **Homo antecessor**

■ **Homo floresiensis**

■ **Homo heidelbergensis**

■ **Homo naledi**

2
Million Years Ago

■ **Homo erectus**

■ **Homo rudolfensis**

■ **Homo habilis**

3
Million Years Ago

4
Million Years Ago

5
Million Years Ago

■ **Ardipithecus ramidus**

6
Million Years Ago

■ **Ardipithecus kadabba**

EARLY HOMININS

■ **Orrorin tugenensis**

7
Million Years Ago

■ **Sahelanthropus tchadensis**

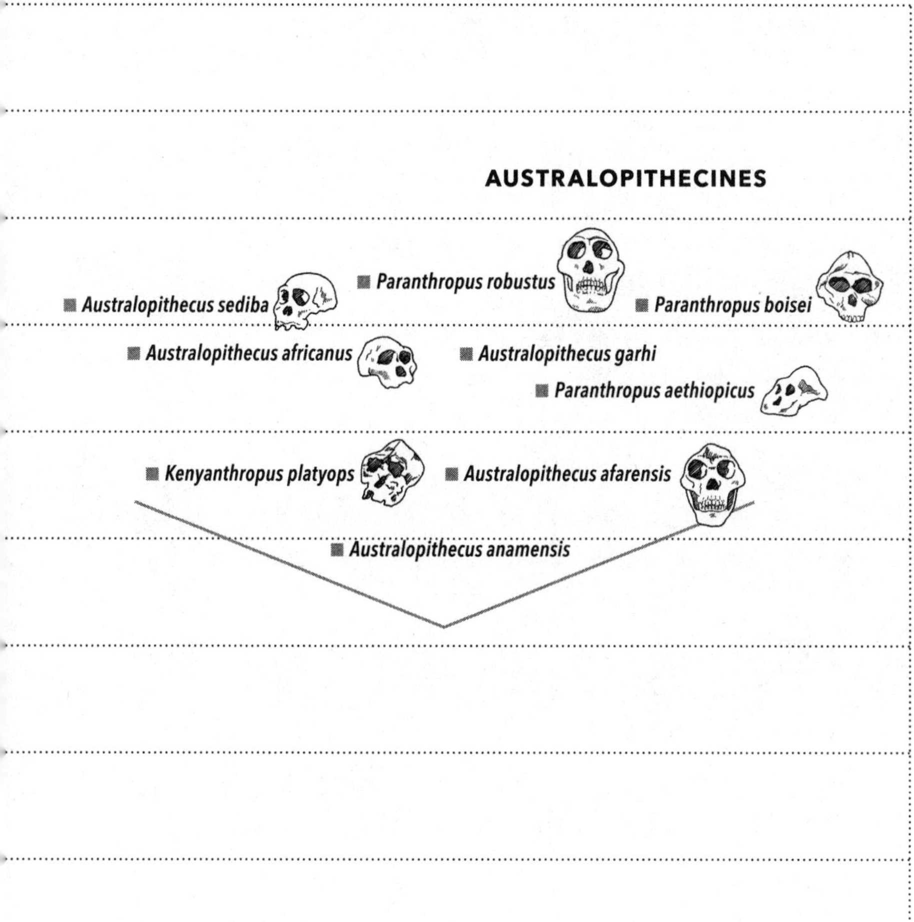

AUSTRALOPITHECINES

Australopithecus sediba

Paranthropus robustus

Paranthropus boisei

Australopithecus africanus

Australopithecus garhi

Paranthropus aethiopicus

Kenyanthropus platyops

Australopithecus afarensis

Australopithecus anamensis

FIGURE 44. Hominin diversity over the past 7 million years; humans are the sole surviving lineage of a once diverse group.

Discovered in 3.2-million-year-old rocks from Ethiopia and named for "Lucy in the Sky with Diamonds," a Beatles tune popular at the time, Lucy was about the size of chimps and Ardi, but had a distinctly larger brain. She still moved gracefully through trees, but her widely spaced hips, arched feet, and short big toe suggest that Lucy walked upright with greater ease than earlier hominins. Lucy's teeth are also distinctive, with large molars well suited for prolonged chewing. Paleoanthropologists posit that Lucy and her kin ate less fruit than chimps and earlier hominins, relying more on tough tubers, seeds, leaves, and stems foraged from open woodlands.

Two additional pieces of evidence illuminate australopithecine biology. In 1976, Mary Leakey discovered a remarkable series of footprints in nearly 3.7-million-year-old rocks from Tanzania. Striding across wet volcanic ash, a man, woman, and child left trackways up to 90 feet (27 meters) long, later buried by more ash. Biologists can tell a lot about your gait from the tracks you leave in the mud, and the Tanzanian footprints show that australopithecines were accomplished walkers, spending much of their day on the ground rather than in trees.

The second bit of evidence is equally remarkable: 3.3-million-year-old rocks from Kenya preserve the oldest known tools, showing that australopithecines (we don't know which ones) fashioned implements by chipping sharp flakes from large, hard stones. In 1957, British anthropologist Kenneth Oakley wrote an influential book titled *Man the Toolmaker*. While other species are known to use nearby objects as simple tools, humans are unique in their ability to design and construct tools for many different purposes. The Kenyan tools, simple as they are, show that the evolutionary pathway that would lead to automobiles, computers, and Frisbees began long before the emergence of our species.

HOMO SAPIENS—us—is the sole extant species of the genus *Homo,* indeed the only living hominin (Figure 44). On the basis of fossils, however, 13 additional species of *Homo* have been recognized (11 of them named formally), all now extinct. Beginning a bit more than 2 million years ago, our closest relatives began to diversify in Africa, much as their hominin forebears did earlier. The best-understood ancestral *Homo* is *Homo erectus,* found in rocks ranging in age from 1.9 million to about 250,000 years ago. Besides its superb preservation based on many individuals, *H. erectus* is notable for two reasons. First, its anatomy is intermediate between those of australopithecines and modern humans, with a more human-like skeleton and a brain larger than Lucy's but smaller than ours. And second, unlike all earlier hominins, *H. erectus* prospered not only in Africa but throughout Eurasia as well. By this time, our ancestors lived firmly on the ground and fed by hunting and gathering food. Cut marks on animal bones show that they butchered their prey, providing an important new source of nutrition as the Earth slid into a full-fledged ice age. Very likely, these ancestors shared their quarry, much as hunter-gatherers do today, promoting social cohesion within the group.

The oldest fossils assigned to *Homo sapiens* occur in 300,000-year-old rocks from Morocco. These fossils are only a bit younger than evidence for a new and sophisticated tool-making culture and the widespread (and controlled) use of fire. So, our species arrived with novel technologies. Perhaps surprisingly, our direct ancestors shared the ice-age Earth with at least three other *Homo* species. Best known are the Neanderthals, often caricatured as brutes, but actually sophisticated hunter-gatherers, with diverse tools and brains larger than our own. At the other end of the spectrum is *Homo florensiensis*, a diminutive relative dubbed "the hobbit," discovered only recently as fossils in Indonesia. And then there are the Denisovians, known only from fragments originally uncovered in 50,000- to 30,000-year-old caves from Siberia and shown to be distinct on the basis of their DNA, preserved in a finger bone. We now have genomes reconstructed from fossils of Neanderthals as well as Denisovians, and these show not only that modern humans, Neanderthals, and Denisovians are closely related but that deep in the past, individuals from these species occasionally interbred. Most people's DNA includes a small admixture of Neanderthal genes; Melanesians, aboriginal Australians, and some other Asian populations also have genes derived from Denisovians. History comes alive in our genes.

Early humans lived only in Africa, but a bit more than 100,000 years ago, one population stuck its toe into the wider world, dwelling in what is now Israel along with Neanderthals. Then, 50,000–70,000 years ago, our species spread rapidly throughout Asia and Europe. What were these intrepid colonists like?

In a windowless room deep within the Museum of Ancient Culture in Tübingen, Germany, small animals, carved from ivory, glisten like jewels (Figure 45). The animals, found in a cave in southwestern Germany, capture their subjects— mammoths, horses, large cats, and more—with remarkable vitality. At 40,000 years, the oldest rank among the earliest known examples of representational art. A nearby cave yielded a female figure, also fashioned from mammoth ivory. About the same age as the oldest animals, it is the earliest known depiction of a human. And across the Old World, the contemporaries of these early carvers began to cover cave walls with exquisite paintings of animals and, perhaps, spirits. The oldest known cave paintings, from Indonesia, were rendered some 44,000 years ago; the hunters depicted on the walls, half-human, half-animal, hint at spirituality as well as art (Figure 46). Tools of this age likewise reflect a new technological revolution, with mass production of stone tools accompanied by finely wrought awls, needles, and even flutes made of bone. Language cannot be inferred from ancient bones, but we might speculate that this key human attribute advanced during this interval, as well. We don't know why these changes happened at this time, but as paleoanthropologist Daniel Lieberman put it, "people were somehow thinking and behaving differently." Humans at last earned the title of modern.

As Plato tells the story (recounted by Adrienne Mayor in her engaging *Gods and Robots*), when the gods created animals, they relegated the task of assigning their capabilities to two titans,

FIGURES 45 AND 46. Humankind's great leap forward: (Figure 45) Exquisite animals carved from mammoth ivory nearly 40,000 years ago. (Figure 46) The oldest known cave paintings, from Indonesia, dating back some 44,000 years. *Figure 45 copyright Museum der Universität Tübingen MUT, J. Lipták; Figure 46 courtesy of Adam Brumm, Griffith University, photo credit Ratno Sardi*

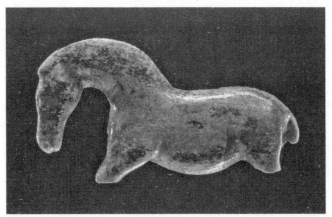

45

46

Prometheus and Epimetheus. Epimetheus, in particular, took to the task with relish, making cheetahs fast, crabs armored, and elephants large. Humans, unfortunately, were last in line, and by the time Epimetheus got around to them, he had distributed all the goodies. Recognizing the urgency of equipping humans for life in the wide world, Prometheus stepped in, providing humans with language, fire, and technology, all stolen from the gods. A lovely story, and actually not far from the view of anthropologists; armed with language, the ability to control fire, and the capacity for making tools, humans began to differentiate themselves from the rest of the animal kingdom.

Without question, changing environments shaped human evolution, as our ancestors adapted to the shifting African landscape. The long-term narrative of Earth's history, however, shows that organisms don't simply reflect their environments. They help to mold them, and in this regard, humans are no different. Well, we are different, but not because we have so little effect on our planet. It's because our influence is so large. *Homo sapiens* has shaped the world around us since our inception and now does so in unprecedented ways, the latest movement in the long symphonic dance of Earth and life.

Twenty thousand years ago, a vast sheet of glacial ice covered the northern half of North America. South of an ice margin that arced from Cape Cod to Montana, tundra, steppe, and spruce forest supported a remarkable diversity of mammals, including mammoths and mastodons, wooly rhinoceroses, cave bears, dire wolves, cave lions and saber-toothed cats, horses, camels, giant sloths, and glyptodons, extinct armadillos the size of a VW Beetle. By 10,000 years ago, all were gone. What happened?

The ice began to melt about 15,000 years ago, and after one last cold snap, Earth warmed rapidly from about 13,000 to 10,000 years ago, introducing the interglacial world we experience today. I call this interval "interglacial" rather than "postglacial" because for the past million years, Earth has oscillated between glacial cold and interglacial warmth on a 100,000-year timescale dictated by metronomic variations in Earth's orbit around the sun. There is no reason to believe that our current warmth is anything other than an interglacial interval destined to give way to renewed glacial advance in the future. At least, there wasn't any reason to think so until industrial humans came along.

As temperate climates expanded in North America 13,000–10,000 years ago, plant populations migrated northward, giving rise to community associations unlike any seen today. A number of scientists have hypothesized that shifting environments and unfamiliar vegetation caused mammal populations to plummet. Environmental stress may indeed have set the stage for mammalian extinctions, but similar climate shifts had occurred repeatedly during the preceding million years without major species loss. Something else seems to have been afoot.

That something was *Homo sapiens*. While humans have a long history in Africa and Eurasia, they didn't arrive in the New World until the waning stages of the last ice age. Recently, archaeologists reported evidence that humans lived along the Salmon River in Idaho 16,500 to 16,300 years ago, recording a first wave of migration from northeastern Asia, probably along the Pacific coast. The human population, whom we call Clovis—after their initial discovery near Clovis, New Mexico— expanded rapidly and gained new and sophisticated stone tools just before the great mammals disappeared. Numerous kill and butcher sites link Clovis culture to hunting, and this, in turn, suggests that humans played a major role in removing large mammal species from North America. In all probability, hunting and environmental change were both involved, but in the absence of humans, the continent's fauna might look different today. In Australia, the arrival of humans 50,000 to 40,000 years ago also coincided with the loss of indigenous animals. By way of contrast, on uninhabited Wrangell Island, an isolated scrap of land in the Chukchi Sea north of Siberia, mammoths survived until about 4,000 years ago. Egyptian pharaohs could have captured them for pageants, had they known where to look.

So humans began to affect the biological Earth early on, and our impact would accelerate through time. A second and ultimately decisive influence began about 11,000 years ago in a crescent that curves northward from Israel and Jordan to Syria, Turkey, and Iraq. Here people first developed agriculture, learning to grow and harvest figs, barley, chickpeas, and lentils. Within a thousand years, sheep, goats, pigs, and cattle had been domesticated. In fact, agriculture developed independently in several parts of the world, including China (9,000 years ago), Meso-America (10,000 years ago), the Andes (7,000 years ago), and parts of sub-Saharan Africa (6,500 years ago). It is currently fashionable to lament this cultural transition, as farming replaced hunting and gathering with a life that entailed more work for a less nutritious and less reliable diet. Perhaps so, but those of us who use iPhones, enjoy movies, or survive cancer might see some advantages to the social reorganization precipitated by the agricultural revolution. Fewer people were needed to produce more food, leaving others to pursue art, invention, and commerce.

Of course, as cropland and grazing expanded, the human

influence on nature grew in proportion. Towns nucleated, and some developed into cites. Population increased, and commerce expanded. That said, the environmental footprint of humans grew slowly at first. Had you lived at the time of Christ or a thousand years later, your life and, indeed, the human impact on our planet would have been broadly similar. Our numbers didn't change much over that interval, hovering around 200 million. As humans learned to exploit the energy resources beneath our feet, however, the trajectories of population size, technological innovation, and environmental influence moved into high gear—in less than two centuries, we went from horse power and steam to gasoline and jet fuel. The human population passed the billion point around 1800, reaching two billion by 1930, and four billion in 1975. We're on course to complete another doubling in the coming decade. And as our population has grown, the environmental impact of each individual has expanded remarkably. Fossil fuels have been extracted since the nineteenth century, but their use has increased nearly tenfold since World War II.

IN SOME WAYS, the Industrial Revolution ushered in a golden age of humanity. Our numbers increased rapidly as the benefits of public health and prosperity expanded widely, if unevenly, around the world. But the very innovations that have allowed us to feed and clothe more than seven billion people now grip the Earth in an increasingly tight vise. Pressure comes from two directions—direct effects on organisms and a mounting influence on Earth's physical environment. A prime example of direct effects, agriculture now takes up half of Earth's habitable surface, displacing plants, animals, and microorganisms that once thrived on these lands. We also challenge natural ecosystems through pollution, affecting air and water, soil and the sea. Of course, pollution exacts a human toll, be it unbreathable air in Delhi or undrinkable water in Flint, Michigan. But it also undermines the diversity, productivity, and ecological resilience of natural communities, with few if any ecosystems unaffected.

To illustrate the point, we can look to the ominously named "dead zones" found in the Gulf of Mexico and other coastal seaways. Throughout mid-continent North America, farmers spread liberal amounts of fertilizer across wheat and corn fields. The fertilizer increases crop yield, but most of its nutrients are never taken up by growing plants; instead, rain and groundwater flush them into rivers, eventually to be disgorged into the Gulf of Mexico. In Gulf waters, the fertilizer finally does its job, promoting seasonal blooms of algae. As the algae sink to the seafloor, they are consumed by respiring bacteria, depleting oxygen in ambient waters. In the absence of the oxygen needed for growth and metabolism, animals on or near the seafloor die in large numbers. In 1988, when the Gulf dead zone was first recognized, it covered an area of 15 square miles (39 square kilometers); in 2017, the dead zone encompassed some 8,776 square miles (27,730 square kilometers), about the size of New Jersey. Several hundred other dead zones have been documented in coastal waters around the world, all toxic to sea life.

We also affect biological diversity directly by selectively exploiting plants and animals for food or commerce, and by transporting species far from their natural homes, where some will invade foreign communities. The rhinoceros, one of our planet's most distinctive and majestic animals, stands as something of a poster child for overexploitation. Prized in parts of Asia for the (fanciful) aphrodisiac properties of their horns, rhinos have long been poached in Africa and Asia. As a result, all rhino populations are endangered, and the northern white Rhino, once found throughout central Africa, is essentially extinct in the wild. Globally, hunting has depleted numerous bird and mammal populations, and many—from condors to elephants—will require active conservation to persist into the world of our grandchildren.

By comparison, the sea strikes most of us as vast and pristine, somehow immune to the depredations of humans. That fiction, however, has been exposed in recent years. We need only look at commercial fisheries to recognize the hand of over-exploitation. Some three billion people depend on seafood for protein, but one in six global fisheries has collapsed in recent decades. Another 30 percent of all commercial fish stocks have been harvested beyond their sustainable limits, and most of the rest have been fished right up to that ecological edge. The collapse of cod stocks on the Grand Banks shows just how badly things can go wrong. Cod populations that yielded more than 800,000 tons of catch in 1958 were declared commercially extinct in 1992, altering the very cultural fabric of adjacent Newfoundland. Commercial fishing was banned, but nearly three decades later, the cod have yet to recover.

Nor does the vastness of the oceans shield them from pollution. It is estimated that about one garbage truck worth of plastic enters the oceans every minute, with a mounting toll on animal populations in many parts of the sea.

HABITAT DISRUPTION, pollution, overexploitation, and invasive species have whittled away at natural ecosystems for more than a century. When we read that more than 10 percent of Australia's indigenous mammal species have disappeared since

European colonization, that North American bird populations have declined by nearly 30 percent since 1970, or that insect populations in European grasslands have dropped by nearly 80 percent over the past decade, the sobering statistics largely reflect these actions. Yet what our grandchildren will recognize as humanity's most profound influence on the Earth is just getting going. As the twenty-first century continues, habitat destruction and the like will not go away, but they will play out on a planet that is, itself, changing dramatically. The big story going forward is global warming, changes to the Earth itself engendered by human participation in the carbon cycle.

To understand this gathering storm, we must turn, one more time, to the fundamental relationship between carbon dioxide and climate, and more broadly to the interactions between Earth and life in cycling carbon. To review, plants and other photosynthetic organisms remove CO_2 from air and water, fixing carbon to form the biomolecules required for growth and reproduction. Animals, fungi, and innumerable microorganisms gain energy by respiring these molecules, returning carbon to the environment as CO_2. Photosynthesis and respiration almost but not quite balance each other, the "not quite" portion being organic matter that escapes respiration and related processes to accumulate in sediments. Some of this buried organic material matures to form petroleum, coal, and natural gas; it will return to the surface carbon cycle only slowly, over millions of years, as plate tectonics lifts the sediments into mountains, exposing them to chemical weathering and erosion. At least that's the way it worked until the Industrial Revolution.

In the physical part of the carbon cycle, volcanoes add CO_2 to the atmosphere, while chemical weathering removes it, the carbon eventually depositing as limestone. Together these processes determine the amount of carbon dioxide in the atmosphere. And because CO_2 is a strong greenhouse gas, they also regulate climate through time. We learned in chapter 7 that 252 million years ago, at the end of the Permian Period, enormous volcanoes delivered massive amounts of CO_2 to the atmosphere, triggering global warming, ocean acidification (a physiologically significant decrease in seawater pH), and oxygen depletion in marine waters. On land and in the sea, biological diversity was devastated. In the warm aftermath of the volcanism, rates of chemical weathering increased, over thousands of years restoring atmospheric CO_2 to its pre-catastrophe level.

Volcanoes may be nature's device for throwing the carbon cycle out of whack, but humans have introduced new and equally potent mechanisms—burning fossil fuels and clearing forests for agriculture. The coal, oil, and natural gas formed over hundreds of millions of years now return their carbon to the atmosphere at fabulous rates. In the twenty-first century, humans contribute 100 times as much carbon dioxide to the atmosphere as all the world's volcanoes put together. But while technological humans have dramatically increased the rate at which CO_2 is added to the atmosphere and oceans, we have not (yet) done anything to increase its rate of removal, so CO_2 in the air around us rises.

Eventually the warmer Earth will increase rates of chemical weathering, rebalancing atmospheric carbon dioxide as it did in the aftermath of end-Permian extinction, but as in the past, that will take thousands of years. From the perspective of our own lives and those of our children and grandchildren, CO_2 is on a one-way trip upward.

We know that atmospheric CO_2 is increasing because we can measure it (Figure 47). In 1958, Charles David Keeling began to monitor the composition of the atmosphere, taking hourly readings from a station atop Mauna Loa, a program that continues today. When Keeling started, the air above Hawaii contained 316 parts per million carbon dioxide. By May 2020, CO_2 had increased to 417 ppm, a value last seen on Earth millions of years ago. In the absence of dramatic societal change, we will reach 500 ppm by mid-century, more like air in the warm world before Antarctic glaciers began than anything experienced by humans or our hominin ancestors.

We know that the observed CO_2 increase is driven mainly by the burning of fossil fuels because they impart a chemical signature to the air. Over the past sixty years, as some scientists measured the *amount* of carbon dioxide in the atmosphere, others measured the carbon-isotopic composition of that CO_2. The ratio of carbon's two stable isotopes, Carbon-12 and Carbon-13, differs among the major carbon reservoirs of the Earth, and we can use these differences to pinpoint the source of CO_2 added to the atmosphere. Carbon dioxide in volcanic gas won't work, and neither will CO_2 dissolved in seawater— their isotopic compositions simply can't explain the changing isotopic composition of atmospheric CO_2. In contrast, organic matter formed via photosynthesis has just the right makeup to explain the data. On the basis of stable isotopes alone, the source of CO_2 added to the atmosphere could be forest clearing or fossil fuels, but when we add analyses of carbon's third isotope, Carbon-14, the answer becomes clear. Because Carbon-14 is radioactive, decaying to nitrogen on a timescale of thousands of years, it is modestly present in living organisms, but undetectable in fossil fuels formed millions of years ago. Mea-

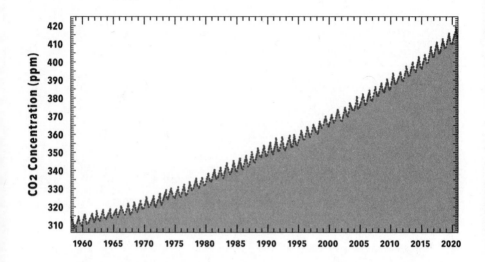

FIGURE 47. The amount of carbon dioxide in the atmosphere, measured hourly since 1958 from a station atop Mauna Loa in Hawaii. The small annual oscillations reflect the fact that there is more land in the Northern Hemisphere than below the equator, and so more photosynthesis in the northern summer, drawing down carbon dioxide levels. In the northern winter, photosynthesis slows but respiration keeps its pace, restoring carbon dioxide to the atmosphere. *Scripps Institution of Oceanography*

surements show that the proportional amount of Carbon-14 in atmospheric CO_2 has declined through time in a way that identifies the principal source of atmospheric CO_2 increase as coal, petroleum, and natural gas, burned by humans to provide energy and warmth for a burgeoning population.

As we add greenhouse gas to the atmosphere, we should expect the Earth's surface to warm, and that is just what is happening—we can measure this, as well (Figure 48). Today we monitor the globe using satellites, but temperatures a century ago must be gleaned from old meteorological and oceanographic records, introducing some uncertainty. Nonetheless, scientific consensus holds that over the past hundred years the mean temperature of our planet's surface has increased by a bit less than 1°C (1.5°F), with the poles warming more rapidly than lower latitudes. The Paris Agreement (2016) among nations pledges the world to limit global warming to less than 2°C above pre-industrial values. We're nearly halfway there already, and although there are tremendous benefits to succeeding, we will fail unless we substantially change our ways.

What are the consequences of a warming Earth? That depends in part on where you are; there will be winners and losers. A recent estimate suggests that by 2050, Toronto will have a climate much like that of present-day Washington, D.C. Some Canadians may look forward to life with less snow, but spare a thought for those who will live in Washington, where in 2050, summer heat and humidity will far exceed the already oppressive weather of today. A Brookings Institution study suggests that in the United States, states along the Canadian border will benefit economically from twenty-first-century climate change, at least a little. Conversely, southern states will pay an economic penalty, in some counties exceeding 15 percent of current income. Some may see poetic justice in this—the largest economic burden falling where climate change denial has been most widespread—but at the end of the day, we will all pay a price as the Earth warms. And as temperature changes, so will precipitation. Water availability is already a geopolitical flashpoint, and as the twenty-first century plays out, it will become ever more important. Decreasing precipitation is forecast for the southwestern United States, populated areas of the Middle East, southwestern Africa, the Iberian Peninsula,

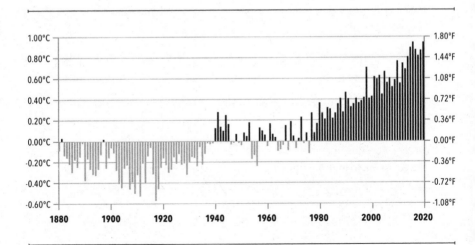

and more. Nearly two billion people who depend on seasonal melt from mountain glaciers in low latitudes will also see water availability dwindle as the glaciers that sustain them shrink and eventually disappear.

Extreme weather, already becoming more frequent, presents another challenge for the twenty-first century and beyond. Devastating fires in California and Australia reflect pronounced warmth and drought, conditions that occurred only rarely in the past. The fear, of course, is that weather extremes will become more common around the world as global change accelerates. The implications for food security and political stability are enormous.

And what about the natural world? How will plants, animals, and microorganisms respond as global change is increasingly layered on habitat disruption, overexploitation, pollution, and species invasions? Populations confronted by changing environments can respond by adaptation, migration (to track their preferred habitat as it shifts), or extinction. Biologists

FIGURE 48. Global temperature over the past 140 years. The chart shows the deviation in May temperatures from their twentieth-century average. Before 1940, global temperatures were consistently below the twentieth-century average; since 1978, they have been consistently above the average and getting warmer by the year. *Source: NOAA Climate.gov*

have documented some compelling examples of rapid adaptation, but the high rates of twenty-first-century global change will challenge many species. Migration will, as well, because in a twenty-first-century world migration routes can be compromised by fields, cities, and superhighways. This being the case, how do we minimize the third option?

National parks, refuges, and other protected lands play a critical role in the conservation of species threatened by habitat disruption and the like. We need the refuges we have and would benefit enormously by their expansion. Yet, how well will we be able to conserve species if the climates of protected areas keep changing? Protected corridors to promote migration will help, but whether lands are protected or not, climate change will alter the distributions of many species. Species that didn't encounter each other in the past will occur in the same place, with largely unknown consequences for competition and ecosystem resilience.

In the face of accelerating climate change, the sea once again seems to show a poker face, its vastness insulating it from human influence. But, again, this perception is dead wrong. For one thing, sea level is rising, as glacial melt flows back into the oceans and warming seawater expands. During the twentieth century mean global sea level increased by 6–8 inches (15–20 centimeters), accelerating in recent years. Estimates for 2100 come with many uncertainties, but most predictions call for an additional rise of 20 to 40 inches (50–100 centimeters). That may not sound like much, but if you live in Venice, Bangladesh, a Pacific atoll—or Florida—sea level change will dramatically alter your life. And as sea levels rise, the physical properties of seawater will change as well. Not surprisingly, as atmospheric CO_2 increases, the oceans, like the land surface, will become warmer. As water warms, it can carry less oxygen gas, so the oceans will lose oxygen, especially at depth. And, insofar as the oceans actually absorb much of the carbon dioxide emitted by human activities, the pH of seawater will decline (ocean acidification). That's right, the trio of killers set in motion by end-Permian volcanism will return in force during the twenty-first century. It's already begun.

As well as anywhere in the world, Australia's Great Barrier Reef epitomizes the intertwined challenges of a changing planet. A remarkable necklace of coral that extends for more than 1,400 miles (2,300 kilometers), the reef has graced the northeastern coast of Australia for millions of years, supporting immense biological diversity while protecting adjacent lands from storms. Despite this long history, a recent study concluded that between 1987 and 2012, the reef lost about 50 percent of its live coral cover, mostly due to cyclones and predation by voracious starfish stimulated by nutrient spikes from agricultural runoff. Now the vise is tightening further, as seawater temperature increases and pH declines. Dozens of laboratory and field experiments show that as seawater pH falls, the ability of corals to secrete their carbonate skeletons decreases. Thus, with ocean acidification accelerating, corals may no longer be able to build the limestone frameworks that define reefs and sustain their biological diversity. And as ocean temperatures rise, another problem comes to the fore. Reef corals are basically farmers, gaining most of their nutrition from algae that live within their tissues. Perhaps surprisingly, when ambient temperature increases beyond a critical point, the corals expel their algae, a process called bleaching because the corals turn white. In the past, when temperature extremes were relatively rare, bleached corals commonly recovered by recruiting more algae. Now, however, rising temperatures make bleaching events more frequent, spelling death for reef corals—in 2016 and 2017, back-to-back bleaching in the northern Great Barrier Reef killed about half the coral colonies in this region, and renewed bleaching in 2020 spread coral loss throughout the reef's remarkable expanse. Intrepid biologists have discovered temperature-tolerant corals in parts of the Pacific Ocean, and these as well as programs of assisted coral recolonization may yet sustain reef ecosystems around the world. But the clock is ticking for some of our planet's most extraordinary ecosystems.

GEOLOGISTS INCREASINGLY REFER to our era as the Anthropocene Epoch, emphasizing the tremendous influence of humans on the world around us and its distinction from all that came before. I think it likely that geologists and paleontologists looking back on our world from the future will recognize the present as unusual, marked by geologically rare rates of environmental change and a reduction of biological diversity similar to those of lesser extinctions in the past, if not (hopefully) the mass extinctions that ended the Paleozoic and Mesozoic eras. Of all the phenomena associated with anthropogenic global change, however, perhaps the most striking is the human response—which, to date, has been slight. It isn't as though we haven't been warned. As early as 1957, oceanographer Roger Revelle spelled out clearly how rising levels of CO_2 in the atmosphere would change climate and, as a consequence, ecosystems around the world. And with each decade since then, the message from scientists has grown clearer—and scarier. It seems difficult for people to get excited by slow changes that play out over decades, but that timescale is misleading. If you're twenty, we're talking about profound changes in your lifetime; if you're sixty, it's the world your grandchildren will face. Fire, hurricanes, water shortages, fisheries collapse, refugee issues— however challenging these seem today, they will become worse as the century proceeds.

There are, of course, those who spread disinformation about global change because they benefit financially from the status quo. The debate about cancer and smoking long ago taught us much about those who prioritize dollars today over a better world tomorrow. Economic arguments for sitting on our hands are both self-serving and specious because they do not take into account the cost of doing nothing. Recent estimates suggest that every dollar spent today to modify the way we live and work will pay a five-dollar dividend by the end of the century.

To be sure, there are uncertainties associated with predictions about future climate and its consequences. The great physicist Niels Bohr is said to have quipped that "it's difficult to make predictions, especially about the future," and whether it was Bohr or someone else who actually said it, the statement is undeniably true. In the past, scientific predictions about twenty-first-century climate change have sometimes been off the mark, but mostly, it turns out, because they have underestimated the pace of change. Scientists are inherently conservative, and we continue to learn about hitherto unappreciated feedbacks that quicken global warming and exacerbate its consequences. Perhaps, then, the best predictions we can make are that (1) "no change" is about the least likely outcome of human activities this century and (2) change may well prove faster and more profound than current models predict.

Dire predictions about the future can instill hopelessness and resignation, but they are actually much like Charles Dickens's Ghost of Christmas Yet to Come in *A Christmas Carol*. The ghost told Scrooge what would come to pass if he did nothing to mend his ways. Scrooge did change, to the benefit of all. To be sure, the challenge of safeguarding our societal future while vouchsafing a natural world shaped by four billion years of evolution is daunting, and each year we do nothing makes the task grow larger and more urgent. Through global commitment, however, we have the capacity to bequeath a safe, sane world to our children. In the developed West, we can decrease our environmental footprint by making wise choices about food, home, and transportation, and we can support sustainable alternatives for those around the world who aspire to better living conditions. As citizens, we can support initiatives to conserve biological diversity and develop Earth-friendly technologies—new forms of batteries (needed to take full advantage of sustainable energy sources) and mechanisms for scrubbing carbon dioxide from the air come readily to mind. In his farewell speech to the American people, George Washington famously warned against "ungenerously throwing upon posterity the burden which we ourselves ought to bear." Washington was speaking of taxes and the national debt, but his words apply equally well to global climate change and its consequences. A generation ago, the United States and its allies focused extraordinary talent and resources on building a bomb; perhaps we can muster the same resolve to provide a better world for our grandchildren.

So here you stand, in the physical and biological legacy of four billion years. You walk where trilobites once skittered across an ancient seafloor, where dinosaurs lumbered across *Gingko*-clad hillsides, where mammoths commanded a frigid plain. Once it was their world, and now it is yours. The difference between you and the dinosaurs, of course, is that you can comprehend the past and envision the future. The world you inherited is not just yours, it is your responsibility. What happens next is up to you.

Acknowledgments

THIS BOOK DISTILLS the fruits of a lifetime spent trying to understand our planet and the life it sustains. Through research on five continents and teaching, first at Oberlin College and then for nearly four decades at Harvard, I have learned a tremendous amount about Earth's past, present, and probable future. In all of these endeavors, I've benefited from the wisdom, collaboration, and support of others.

Scientists generally stand at the confluence of two intellectual streams. First, there is all that flowed to us from our own teachers. My mentors included Elso Barghoorn, pioneer in the paleontological search for Earth's earliest life; Dick Holland, a towering geochemist who set the stage for research on Earth's environmental history; Stephen Jay Gould, who fueled my interest in evolution; Ray Siever, who encouraged me to look carefully at sedimentary rocks; and Steve Golubic, who taught me about cyanobacteria. The other stream connects us with the students and postdocs who have worked in our laboratories—a steady stream of ideas and insights that definitely runs in two directions. Knoll lab alumni are a superb group of scientists who are taking studies of paleontology, geobiology, and Earth history in new directions, and I am grateful for and proud of them all.

The list of coauthors on my scientific papers through the years runs to more than five hundred, and I can't mention everyone here, although I appreciate them all. I must, however, thank John Hayes, who taught me everything I know about biogeochemistry; Keene Swett and Brian Harland, who introduced me to arctic research; Malcolm Walter, friend and fieldmate in numerous forays into the Australian Outback; Misha Semikhatov and Volodya Sergeev, comrades in the geological exploration of Siberia; Mario Giordano, who translated my paleontological hunches into laboratory experiments; John Grotzinger, partner for the past thirty years in fieldwork that has ranged from Namibia and Siberia to (virtually at least) Mars; and Dick Bambach, who has long challenged me to think about evolution in new ways.

Ernest Hemingway had Maxwell Perkins to help shape his novels, and fortunately I have Peter Hubbard. *A Brief History of Earth* was Peter's idea, and his support, advice, and constructive criticism have lifted every page of the book. I also thank Molly Gendell and indeed everyone at HarperCollins, professionals all. And for graciously sharing some of the images used in this book, I thank the Atacama Large Millimeter Array, Matteo Chinellato (via Wiki, Creative Commons), Marie Tharp Maps LLC and Lamont-Doherty Earth Observatory, Ron Blakey of Deep-Time Maps, the Smithsonian Institution's National Museum of Natural History, the American Museum of Natural History, the Museum of Ancient Cultures at the Eberhardt Karls University of Tübingen, the Scripps Institute of Oceanography, and the National Oceanographic and Atmospheric Administration, as well as my friends and colleagues Zhu Maoyan, Nick Butterfield, Shuhai Xiao, Guy Narbonne, Mansi Srivastava, Frankie Dunn, Alex Liu, Misha Fedonkin, Jean-Bernanrd Caron, Alex Brasier, Hans Kerp, Hans Steur, Neil Shubin, Mike Novacek, and Adam Brumm.

Last, and most important, I thank my home team: Marsha, Kirsten, and Rob. Without their love and support, this book (and much else) wouldn't exist.

Further Reading

1 | CHEMICAL EARTH

Approachable Readings

Eric Chaisson (2006). *Epic of Evolution: Seven Ages of the Cosmos.* Columbia University Press, New York, 478 pp.

Robert M. Hazen (2012). *The Story of Earth: The First 4.5 Billion Years, from Stardust to Living Planet.* Viking, New York, 306 pp.

Harry Y. McSween (1997). *Fanfare for Earth: The Origin of Our Planet and Life.* St. Martin's Press, New York, 252 pp.

Neil de Grasse Tyson (2017). *Astrophysics for People in a Hurry.* W. W. Norton and Company, New York, 222 pp.

More Technical References

Edwin Bergin and others (2015). "Tracing the Ingredients for a Habitable Earth from Interstellar Space Through Planet Formation." *Proceedings of the National Academy of Sciences, USA* 112: 8965–8970.

T. Mark Harrison (2009). "The Hadean Crust: Evidence from >4 Ga Zircons." *Annual Review of Earth and Planetary Sciences* 37: 479–505.

Roger H. Hewins (1997). "Chondrules." *Annual Review of Earth and Planetary Sciences* 25: 61–83.

Anders Johansen and Michiel Lambrechts (2017). "Forming Planets via Pebble Accretion." *Annual Review of Earth and Planetary Sciences* 45: 359–87.

Harold Levison and others (2015). "Growing the Terrestrial Planets from the Gradual Accumulation of Submeter-sized Objects." *Proceedings of the National Academy of Sciences, USA* 112: 14180–85.

Bernard Marty (2012). "The Origins and Concentrations of Water, Carbon, Nitrogen and Noble Gases on Earth." *Earth and Planetary Science Letters* 313–14: 56–66.

Anne Peslier (2020). "The Origins of Water." *Science* 369: 1058.

Laurette Piani and others (2020). "Earth's Water May Have Been Inherited from Material Similar to Enstatite Chondrite Meteorites." *Science* 369: 1110–13.

Elizabeth Vangioni and Michel Cassé (2018). "Cosmic Origin of the Chemical Elements Rarety in Nuclear Astrophysics." *Frontiers in Life Science* 10: 84–97.

Jonathan P. Williams and Lucas A. Cieza (2011). "Protoplanetary Disks and Their Evolution." *Annual Review of Astronomy and Astrophysics* 49: 67–117.

Kevin Zahnle (2006). "Earth's Earliest Atmosphere." *Elements* 2: 217–22.

2 | PHYSICAL EARTH

Approachable Readings

Charles H. Langmuir and Wally Broecker (2012). *How to Build a Habitable Planet: The Story of Earth from the Big Bang to Humankind.* Princeton University Press, Princeton, NJ, 736 pp.

Alan McKirdy and others (2017). *Land of Mountain and Flood: The Geology and Landforms of Scotland.* None Edition, Birlinn Ltd., Edinburgh, Scotland, 322 pp. (This is an informative travel guide to Scotland; Mountain Press publishes a series of Roadside Geology books for curious travelers in the United States.)

Naomi Oreskes, editor (2003). *Plate Tectonics: An Insider's History of the Modern Theory of the Earth.* Westview Press, Boulder, CO, 448 pp. (republished as an ebook in 2018 by the CRC Press).

United States Geological Survey, website: "Understanding Plate Motions." https://pubs.usgs.gov/gip/dynamic/understanding.html.

More Technical References

Annie Bauer and others (2020). "Hafnium Isotopes in Zircons Document the Gradual Onset of Mobile-lid Tectonics." *Geochemical Perspectives Letters* 14: 1–6.

Jean Bédard (2018). "Stagnant Lids and Mantle Overturns: Implications for Archaean Tectonics, Magmagenesis, Crustal Growth, Mantle Evolution, and the Start of Plate Tectonics." *Geoscience Frontiers* 9: 19–49.

Ilya Bindeman and others (2018). "Rapid Emergence of Subaerial Landmasses and Onset of a Modern Hydrologic Cycle 2.5 Billion Years Ago." *Nature* 557: 545–48.

Alec Brenner and others (2020). "Paleomagnetic Evidence for Modern-like Plate Motion Velocities at 3.2 Ga." *Science Advances* 6, no. 17, eaaz8670, doi:10.1126/sciadv.aaz8670.

Peter Cawood and others (2018). "Geological Archive of the Onset of Plate Tectonics." *Philosophical Transactions of the Royal Society,* London. 376A: 20170405, doi: 10.1098/rsta.20170405.

Chris Hawkesworth and others (2020). "The Evolution of the Continental Crust and the Onset of Plate Tectonics." *Frontiers in Earth Science* 8: 326, doi: 10.3389/feart.2020.00326.

Anthony Kemp (2018). "Early Earth Geodynamics: Cross Examining the Geological Testimony." *Philosophical Transactions of the Royal Society,* London. 371A: 20180169, doi: 10.1098/rsta.2018.0169.

Jun Korenaga (2013). "Initiation and Evolution of Plate Tectonics on Earth: Theories and Observations." *Annual Review of Earth and Planetary Sciences* 41: 117–51.

Craig O'Neill and others (2018). "The Inception of Plate Tectonics: A Record of Failure." *Philosophical Transactions of the Royal Society,* London. 371A: 20170414, doi: 10.1098/rsta.20170414.

3 | BIOLOGICAL EARTH

Approachable Readings

David Deamer (2019). *Assembling Life: How Can Life Begin on Earth and Other Habitable Planets?* Oxford University Press, Oxford, UK, 184 pp.

Paul G. Falkowski (2015). *Life's Engines: How Microbes Made Earth Habitable.* Princeton University Press, Princeton, NJ, 205 pp.

Andrew H. Knoll (2003). *Life on a Young Planet: The First Three Billion Years of Life on Earth.* Princeton University Press, Princeton, NJ, 277 pp.

Nick Lane (2015). *The Vital Question: Energy, Evolution and the Origins of Complex Life.* W. W. Norton and Company, New York, 360 pp.

Martin Rudwick (2014). *Earth's Deep History: How It Was Discovered and Why It Matters.* University of Chicago Press, Chicago, 360 pp.

More Technical References

Abigail Allwood and others (2006). "Stromatolite Reef from the Early Archaean Era of Australia." *Nature* 441: 714–18.

Giada Arney and others (2016). "The Pale Orange Dot: The Spectrum and Habitability of Hazy Archean Earth." *Astrobiology* 16: 873–99.

Tanja Bosak and others (2013). "The Meaning of Stromatolites." *Annual Review of Earth and Planetary Sciences* 41: 21–44.

Martin Homann (2018). "Earliest Life on Earth: Evidence from the Barberton Greenstone Belt, South Africa." *Earth-Science Reviews* 196, doi: 10.1016/j.earscirev.2019.102888.

Emmanualle Javaux (2019). "Challenges in Evidencing the Earliest Traces of Life." *Nature* 572: 451–60.

Gerald Joyce and Jack Szostak (2018). "Protocells and RNA Self-replication." *Cold Spring Harbor Perspectives in Biology,* doi: 10.1101/cshperspect.a034801.

William Martin (2020). "Older Than Genes: The Acetyl CoA Pathway and Origins." *Frontiers in Microbiology* 11: 817, doi: 10.3389/fmicb.2020.00817.

Matthew Powner and John Sutherland (2011). "Prebiotic Chemistry: A New Modus Operandi." *Philosophical Transactions of the Royal Society,* London. 366B: 2870–77.

Alonso Ricardo and Jack Szostak (2009). "Origins of Life on Earth." *Scientific American* 301, no. 3, Special Issue: 54–61.

Eric Smith and Harold Morowitz (2016). *The Origin and Nature of Life on Earth: The Emergence of the Fourth Geosphere.* Cambridge University Press, Cambridge, UK, 691 pp.

Norman Sleep (2018). "Geological and Geochemical Constraints on the Origin and Evolution of Life." *Astrobiology* 18: 1199–1219.

4 | OXYGEN EARTH

Approachable Readings

John Archibald (2014). *One Plus One Equals One.* Oxford University Press, Oxford, UK, 205 pp.

Donald E. Canfield (2014). *Oxygen: A Four Billion Year History.* Princeton University Press, Princeton, NJ, 196 pp.

Nick Lane (revised edition, 2016). *Oxygen: The Molecule That Made the World.* Oxford University Press, Oxford, UK, 384 pp.

More Technical References

Ariel Anbar and others (2007). "A Whiff of Oxygen Before the Great Oxidation Event?" *Science* 317: 1903–6.

Andre Bekker and others (2010). "Iron Formation: The Sedimentary Product of a Complex Interplay Among Mantle, Tectonic, Oceanic, and Biospheric Processes." *Economic Geology* 105: 467–508.

David Catling (2014). "The Great Oxidation Event Transition." *Treatise on Geochemistry* (second edition) 6: 177–95.

T. Martin Embley and William Martin (2006). "Eukaryotic Evolution, Changes and Challenges." *Nature* 440: 623–30.

Laura Eme and others (2017). "Archaea and the Origin of Eukaryotes." *Nature Reviews in Microbiology* 15: 711–23.

Jihua Hao and others (2020). "Cycling Phosphorus on the Archean Earth: Part II. Phosphorus Limitation on Primary Production in Archean Oceans." *Geochimica et Cosmochimica Acta* 280: 360–77.

Heinrich Holland (2006). "The Oxygenation of the Atmosphere and Oceans." *Philosophical Transactions of the Royal Society,* London. 361B: 903–15.

Olivia Judson (2017). "The Energy Expansions of Evolution." *Nature Ecology and Evolution* 1: 138.

Andrew H. Knoll and others (2006). "Eukaryotic Organisms in Proterozoic Oceans." *Philosophical Transactions of the Royal Society,* London. 361B: 1023–38.

Timothy Lyons and others (2014). "The Rise of Oxygen in Earth's Early Ocean and Atmosphere." *Nature* 506: 307–15.

Simon Poulton and Donald Canfield (2011). "Ferruginous Conditions: A Dominant Feature of the Ocean Through Earth's History." *Elements* 7: 107–12.

Jason Raymond and Daniel Segre (2006). "The Effect of Oxygen on Biochemical Networks and the Evolution of Complex Life." *Science* 311: 1764–67.

Bettina Schirrmeister and others (2016). "Cyanobacterial Evolution During the Precambrian." *International Journal of Astrobiology* 15: 187–204.

5 | ANIMAL EARTH

Approachable Readings

Mikhail Fedonkin and others (2007). *The Rise of Animals: Evolution and Diversification of the Kingdom Animalia.* Johns Hopkins University Press, Baltimore, MD, 344 pp.

Richard Fortey (2001). *Trilobite; Eyewitness to Evolution.* Vintage, New York, 320 pp.

John Foster (2014). *Cambrian Ocean World: Ancient Sea Life of North America.* Indiana University Press, Bloomington, IN, 416 pp.

Stephen Jay Gould (1990). *Wonderful Life: The Burgess Shale and the Nature of History.* W. W. Norton and Company, New York, 352 pp.

More Technical References

Graham Budd and Sören Jensen (2000). "A Critical Reappraisal of the Fossil Record of the Bilaterian Phyla." *Biological Reviews* 75: 253–95.

Allison Daley and others (2018). "Early Fossil Record of Euarthropoda and the Cambrian Explosion." *Proceedings of the National Academy of Sciences, USA* 115: 5323–31.

Patricia Dove (2010). "The Rise of Skeletal Biominerals." *Elements* 6: 37–42.

Douglas Erwin and James Valentine (2013). *The Cambrian Explosion: The Construction of Animal Biodiversity.* W. H. Freeman, New York, 416 pp.

Douglas Erwin and others (2011). "The Cambrian Conundrum: Early Divergence and Later Ecological Success in the Early History of Animals." *Science* 334: 1091–97.

P.U.P.A. Gilbert and others (2019). "Biomineralization by Particle Attachment in Early Animals." *Proceedings of the National Academy of Sciences, USA* 116: 17659–65.

Paul Hoffman (2009). "Neoproterozoic Glaciation." *Geology Today* 25: 107–14.

Andrew H. Knoll (2011). "The Multiple Origins of Complex Multi-cellularity." *Annual Review of Earth and Planetary Sciences* 39: 217–39.

M. Gabriela Mángano and Luis Buatois (2020). "The Rise and Early Evolution of Animals: Where Do We Stand from a Trace-Fossil Perspective?" *Interface Focus* 10, no. 4: 20190103.

Guy Narbonne (2005). "The Ediacara Biota: Neoproterozoic Origin of Animals and Their Ecosystems." *Annual Review of Earth and Planetary Sciences* 33: 421–42.

Erik Sperling and Richard Stockey (2018). "The Temporal and Environmental Context of Early Animal Evolution: Considering All the Ingredients of an 'Explosion.'" *Integrative and Comparative Biology* 58: 605–22.

Alycia Stigall and others (2019). "Coordinated Biotic and Abiotic Change During the Great Ordovician Biodiversification Event: Darriwilian Assembly of Early Paleozoic Building Blocks." *Palaeogeography, Palaeoclimatology, Palaeoecology* 530: 249–70.

Shuhai Xiao and Marc Laflamme (2008). "On the Eve of Animal Radiation: Phylogeny, Ecology and Evolution of the Ediacara Biota." *Trends in Ecology and Evolution* 24: 31–40.

6 | GREEN EARTH

Approachable Readings

Steve Brusatte (2018). *The Rise and Fall of the Dinosaurs: A New History of a Lost World*. HarperCollins, New York, 404 pp.

Paul Kenrick (2020). *A History of Plants in Fifty Fossils*. Smithsonian Books, Washington, D.C., 160 pp.

Neil Shubin (2008). *Your Inner Fish: A Journey into the 3.5-Billion-Year History of the Human Body*. Pantheon Books, New York, 229 pp.

More Technical References

Jennifer Clack (2012). *Gaining Ground: The Origin and Evolution of Tetrapods*. Second edition. Indiana University Press, Bloomington, IN, 544 pp.

Blake Dickson and others (2020). "Functional Adaptive Landscapes Predict Terrestrial Capacity at the Origin of Limbs." *Nature*: doi.org/10.1038/s41586-020-2974-5.

Else Marie Friis and others (2011). *Early Flowers and Angiosperm Evolution*. Cambridge University Press, Cambridge, UK, 595 pp.

Patricia Gensel (2008). "The Earliest Land Plants." *Annual Review of Ecology, Evolution and Systematics* 39: 459–77.

Patrick Herendeen and others (2017). "Palaeobotanical Redux: Revisiting the Age of the Angiosperms." *Nature Plants* 3: 17015, doi: 10.1038/nplants.2017.15.

Zhe-Xi Luo (2007). "Transformation and Diversification in Early Mammal Evolution." *Nature* 450: 1011–19.

Jennifer Morris and others (2018). "The Timescale of Early Land Plant Evolution." *Proceedings of the National Academy of Sciences, USA* 115: E2274–83.

Eoin O'Gorman and and David Hone (2012). "Body Size Distribution of the Dinosaurs." *PLOS One* 7(12): e51925.

Jack O'Malley-James and Lisa Kaltenegger (2018). "The Vegetation Red Edge Biosignature Through Time on Earth and Exoplanets." *Astrobiology* 18: 1127–36.

P. Martin Sander and others (2011). "Biology of the Sauropod Dinosaurs: the Evolution of Gigantism." *Biological Reviews* 86: 117–55.

Chistine Strullu-Derrien and others (2019). "The Rhynie Chert." *Current Biology* 29: R1218–23.

7 | CATASTROPHIC EARTH

Approachable Readings

Walter Alvarez (updated edition, 2015). *T. rex and the Crater of Doom.* Princeton University Press, Princeton, NJ, 208 pp.

Michael Benton (2005). *When Life Nearly Died: The Greatest Mass Extinction of All Time.* Thames & Hudson, London, 336 pp.

Douglas Erwin (updated edition, 2015). *Extinction: How Life on Earth Nearly Ended 250 Million Years Ago.* Princeton University Press, Princeton, NJ, 320 pp.

More Technical References

Luis W. Alvarez and others (1980). "Extraterrestrial Cause for the Cretaceous-tertiary Extinction." *Science* 208: 1095–108.

Richard K. Bambach (2006). "Phanerozoic Biodiversity: Mass Extinctions." *Annual Review of Earth and Planetary Sciences* 34: 127–55.

Richard K. Bambach and others (2004). "Origination, Extinction, and Mass Depletions of Marine Diversity." *Paleobiology* 30: 522–42.

Seth Burgess and others (2014). "High-precision Timeline for Earth's Most Severe Extinction." *Procedings of the National Academy of Sciences, USA* 111: 3316–21.

Jacopo Dal Corso and others (2020). "Extinction and Dawn of the Modern World in the Carnian (Late Triassic)." *Science Advances* 6: eaba0099.

Seth Finnegan and others (2012). "Climate Change and the Selective Signature of the Late Ordovician Mass Extinction." *Proceedings of the National Academy of Sciences, USA* 109: 6829–34.

Sarah Greene and others (2012). "Recognising Ocean Acidification in Deep Time: An Evaluation of the Evidence for Acidification Across the Triassic-Jurassic Boundary." *Earth-Science Reviews* 113: 72–93.

Pincelli Hull and others (2020). "On Impact and Volcanism Across the Cretaceous-Paleogene Boundary." *Science* 367: 266–72.

Wolfgang Kiessling and others (2007). "Extinction Trajectories of Benthic Organisms Across the Triassic–Jurassic Boundary." *Palaeogeography, Palaeoclimatology, Palaeoecology* 244: 201–22.

Andrew H. Knoll and others (2007). "A Paleophysiological Perspective on the End-Permian Mass Extinction and Its Aftermath." *Earth and Planetary Science Letters* 256: 295–313.

Jonathan L. Payne and Matthew E. Clapham (2012). "End-Permian Mass Extinction in the Oceans: An Ancient Analog for the Twenty-First Century?" *Annual Review of Earth and Planetary Science* 40: 89–111.

Bas van de Schootbrugge and Paul Wignall (2016). "A Tale of Two Extinctions: Converging End-Permian and End-Triassic Scenarios." *Geological Magazine* 153: 332–54.

Peter Schulte and others (2010). "The Chicxulub Asteroid Impact and Mass Extinction at the Cretaceous-Paleogene Boundary." *Science* 327: 1214–18.

8 | HUMAN EARTH

Approachable Readings

Sandra Diaz and others, editors (2019). Intergovernmental Science-Policy Platform on Biodiversity and Ecosystem Services (IPBES), Summary for Policymakers of the Global Assessment Report of the Intergovernmental Science-Policy Platform on Biodiversity and Ecosystem Services. IPBES Secretariat. https://ipbes.net/global-assessment-report-biodiversity-ecosystem-services.

Yuval Noah Harari (2015). *Sapiens: A Brief History of Humankind.* HarperCollins, New York, 443 pp.

Louise Humphrey and Chris Stringer (2019). *Our Human Story.* Natural History Museum, London, 158 pp.

Elizabeth Kolbert (2014). *The Sixth Extinction: An Unnatural History.* Henry Holt and Company, New York, 319 pp.

Daniel Lieberman (2013). *The Story of the Human Body: Evolution, Health and Disease.* Vintage, New York, 460 pp.

Mark Muro and others (2019). "How the Geography of Climate Damage Could Make the Politics Less Polarizing." Brookings Institution Report; https://www.brookings.edu/research/how-the-geography-of-climate-damage-could-make-the-politics-less-polarizing. (See also *The Economist,* September 21–27, 2019, pp. 31–32.)

Callum Roberts (2007). *The Unnatural History of the Sea.* Island Press, Washington, D.C., 435 pp.

More Technical References

Jean-Francois Bastin and others (2019). "Understanding Climate Change from a Global Analysis of City Analogues." *PLOS One* 14(7): e0217592.

Glenn De'ath and others (2012). "The 27-Year Decline of Coral Cover on the Great Barrier Reef and Its Causes." *Proceedings of the National Academy of Sciences, USA* 109: 17995–99.

Sandra Diaz and others (2019). "Pervasive Human-driven Decline of Life on Earth Point to the Need for Transformative Change." *Science* 366: eaax3100, doi: 10.1126/science.aaw3100.

Rudolfo Dirzo and others (2014). "Defaunation in the Anthropocene." *Science* 345: 401–6.

Jacquelyn Gill and others (2011). "Pleistocene Megafaunal Collapse, Novel Plant Communities, and Enhanced Fire Regimes in North America." *Science* 326: 1100–103.

Peter Grant and others (2017). "Evolution Caused by Extreme Events." *Philosophical Transactions of the Royal Society*, London. 372B: 20160146.

Ove Hoegh-Guldberg and others (2019). "The Human Imperative of Stabilizing Global Climate Change at 1.5°C." *Science* 365: eaaw6974.

Paul Koch and Anthony Barnosky (2006). "Late Quaternary Extinctions: State of the Debate." *Annual Review of Ecology Evolution and Systematics* 37: 215–50.

Xijun Ni and others (2013). "The Oldest Known Primate Skeleton and Early Haptorhine Evolution." *Nature* 498: 60–64.

Bernhart Owen and others (2018). "Progressive Aridification in East Africa over the Last Half Million Years and Implications for Human Evolution." *Proceedings of the National Academy of Sciences, USA* 115: 11174–79.

Felisa Smith and others (2019). "The Accelerating Influence of Humans on Mammalian Macroecological Patterns over the Late Quaternary." *Quaternary Science Reviews* 211: 1–16.

John Woinarski and others (2015). "Ongoing Unraveling of a Continental Fauna: Decline and Extinction of Australian Mammals Since European Settlement." *Proceedings of the National Academy of Sciences, USA* 112: 4531–40.

Bernard Wood (2017). "Evolution: Origin(s) of Modern Humans." *Current Biology* 27: R746–69.

Index